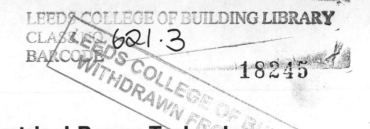
Electrical Power Technology

Electrical Power Technology

David W. Tyler

BH NEWNES

Newnes
An imprint of Butterworth-Heinemann
Linacre House, Jordan Hill, Oxford OX2 8DP
A division of Reed Educational and Professional Publishing Ltd

 A member of the Reed Elsevier plc group

OXFORD   BOSTON   JOHANNESBURG
MELBOURNE   NEW DELHI   SINGAPORE

First published 1997

British Library Cataloguing in Publication Data
Tyler, D. W. (David W.)
  Electrical power technology
  1. Electric engineering
  I. Title
  621.3

ISBN 0 7506 3470 7

Typeset by Laser Words, Madras, India
Printed and bound in Great Britain by the Bath Press, Bath

Contents

Preface

This book has been written for the Advanced GNVQ in Engineering, covering the BTEC Optional Unit 12, Electrical Power Technology, and the City & Guilds Optional Unit 19, Electrical Power. It also covers the BTEC NII level objectives in Electrical Applications U86/330. To understand the applications, a knowledge of the underlying principles is needed and these are covered briefly in the text.

Throughout each chapter are worked examples and **Test your knowledge** questions. When reached, the examples should be worked through and the questions attempted before moving on. The **Test your knowledge** questions are mostly very short and require information given up to that point in the chapter. They enable you to check whether you have learnt the salient points of the preceding work. Answers to these questions are given at the end of the book.

At various stages in the chapters there are **Activities** which will require much more effort to accomplish. These will create work suitable for a portfolio of evidence, and also provide opportunities to develop Key Skills in Communications, Information Technology and Application of Number. It is not envisaged that all students will tackle all of the **Activities**; which ones are attempted will be determined at least in part by the equipment you have at your disposal.

At the end of each chapter there is a further selection of problems given with mathematical answers in brackets. Answers to questions requiring descriptive answers will be found by looking back through the text. Concluding each chapter, a section of **Multiple choice questions** provides practice material for GNVQ end-of-unit tests.

In a subject such as this, many pieces of work require you to describe operations or effects. A good diagram is worth many words and this book aims to promote the use of diagrams by the inclusion of over 200 throughout the nine chapters.

I wish to thank the Institution of Electrical Engineers for permission to quote from the Regulations for Electrical Installations. Any interpretation placed on these regulations is mine alone.

In these days of ever-increasing use of computers, there is a growing tendency to regard any piece of equipment without a screen as being in some way deficient. I would like to remind those who have this impression that, without power equipment and 'power men', the rest of the industry would not exist.

I wish all my readers success in their studies and hope that all, including those who are not following a formal course, will find the material in this book useful, interesting, and perhaps most of all, approachable.

David W. Tyler

1 Electricity supply

Summary

This chapter looks at energy sources for generation of electrical energy and the basic cycle employed. It explains why transmission is carried out at very high voltages and shows how the various pieces of equipment are interconnected using switchgear and transformers to give maximum efficiency together with practical user voltages. It compares the use of underground cables with overhead lines. The purpose of switchgear is described together with the equipment found in distribution substations. The reasons for the adoption of the three-phase system is explained and the losses in feeder systems examined.

Synchronous generators

Virtually all the generation of electrical energy throughout the world is done using three phase synchronous generators. Almost invariably the synchronous generator has its magnetic field produced electrically by passing direct current through a winding on an iron core which rotates between the three windings or phases of the machine. These windings are embedded in slots in an iron stator and one end of each winding is connected to a common point and earthed. The output from the generator is taken from the other three ends of the windings. The output from a three-phase generator is therefore carried on three wires. In many three-phase diagrams single line representation is used when each line on the diagram represents three identical conductors. Figure 1.2 is drawn using this method.

All such generators connected to a single system must rotate at exactly the same speed, hence the description *synchronous* generator.

They are driven by prime movers using steam generated by burning coal or oil, by nuclear reactors, water falling from a higher to a lower level, or aircraft-type gas turbines burning oil or gas. A very small amount of generation is carried out using diesel engines.

Generators range in size from 70 MVA (60 MW at 0.85 power factor) at a line voltage of 11 kV which were mostly installed in the 1950s, through an intermediate size of 235 MVA (200 MW at 0.85 power factor) to more recent machines of 660 and 1000 MW which generate at 25.6 kV. The very large machines are quite rare; many modern power stations employ a combined cycle where gas turbines running on gas produce electricity, and the extremely hot exhaust gases raise steam in a boiler which is used in a further turbine to generate more power. Such combined cycle stations have an efficiency about 50 per cent greater than a straight steam power station but individually are of lower rating.

Energy sources

- **Coal.** Coal is plentiful worldwide and there are sufficient known reserves to last for centuries. It may be deep mined or extracted from open cast mines. Open cast coal is generally of lower quality than deep mined coal but is more easily extracted since once the top overburden is removed it can be scraped out by very large diggers. Open cast mining is opposed on environmental grounds because of the huge areas of land involved and the dust produced. When coal is burnt it produces sulphur oxides which combine with water to form sulphurous and sulphuric acids which are very corrosive. The term *acid rain* is used in this context. Large amounts of ash are produced which need to be disposed of, filling in quarries and marshes etc. To plan and build a coal-fired power station takes several years and it must be accessible to bulk transport to bring in the coal (5 million tonnes annually typically for a major station) and to get rid of the ash which may be up to one tenth of the coal input.

- **Oil.** This may be residual oil from refineries or refined products.
  - (a) Residual oil. When oil from the well has been refined and the petrol, paraffin and diesel oil taken off, the refinery is left with a tar-like substance which is only liquid when kept hot. This may be burnt in power stations to produce steam in the same manner as coal. It should be cheap since the refineries need to get rid of it but is sometimes costed as coal equivalent for heat production. It contains all the impurities of the original raw oil and produces large amounts of sulphur oxides but little ash. It needs special transport to keep it hot and must be maintained hot at the power station since once cold it cannot be pumped.
  - (b) Refined oil. This may be kerosene for use in gas turbines or diesel oil for use in reciprocating engines. This is a very clean fuel but relatively expensive. It burns mainly to water vapour and carbon dioxide although in common with all high-temperature processes there will be some nitrogen oxides produced which are also associated with acid rain. Stations burning refined fuel are generally much simpler to construct than those involving solid fuels.

- **Natural gas.** This is comparable to refined oil but is richer in hydrogen so that when it burns it produces more water vapour and less carbon dioxide and is therefore favoured by those concerned with the effects of carbon dioxide on the global atmosphere (global warming). Stations using refined fuel are quick to start and shut down and so are often used for supplying short peak demands.

- **Nuclear.** Heat produced by nuclear fission of uranium derivatives is used to produce steam and the cycle then continues as with burning any other fuel in a boiler. Nuclear power stations are immensely expensive to plan, construct, fuel up the initial charge and to bring on line. However, once running they produce electricity at virtually zero incremental cost, the main charge being paying off the investment. Therefore nuclear stations are ideally suited to run on base load, maintaining their output 24 hours a day for months at a time. The principal disadvantage will probably turn out to be the cost of decommissioning the stations at the end of their life. Once the turbines and external pipework have been removed the reactors will need to be boxed up and kept secure for centuries before anyone can finally dismantle them.

- **Water.** Very large quantities of water falling through a short distance can be used to drive Francis-type water turbines connected to generators. Again, smaller quantities of water falling through perhaps several hundred metres can be used with Pelton Wheel turbines to the same end. In Britain the possibilities are quite small; there are hydroelectric stations in Scotland and in Wales but their outputs are small compared with fuel-fired stations. There are some very large installations in other countries, a very famous one being the Hoover dam on the Colorado river in the USA.

- **Wind.** In situations where strong winds can be guaranteed for long periods wind turbines are being installed. Generally they involve a two- or three-bladed rotor driving an alternator rated at up to 500 kW set on a single tall pillar. These are often arranged in groups of 10 to 15. Once running they have zero fuel input charge, the whole of their output producing revenue which has to pay off the original investment.

- **Refuse, sewage sludge, peat, wood chippings, straw, vehicle tyres, garbage.** Refuse tips produce large quantities of methane gas which has caused problems when it leaked into houses causing potentially explosive situations. This can be piped away and allowed to burn off uselessly but there are now many instances where it is used to drive diesel-type engines with some oil backup to produce electricity. The same is true of gases produced from sewage sludge. The power produced is used to supply the sewage works, any excess being exported to the system. Willow wood grows very quickly and in very wet areas, Ireland typically, this can be coppiced, chipped and then burnt as coal in specially designed boilers to produce steam and electricity. Peat is also used. Now that farmers are no longer allowed to burn straw in the fields in Britain there is a move to use this as a fuel. Boilers are available which burn whole bales, several at a time, but of course the ratings are fairly small. Huge numbers of vehicle tyres need to be disposed of annually and plants are being developed to burn these to produce electricity and steel scrap from the reinforcement at the same time. Household garbage is also burnt to produce power in some cities.

Generation cycles

Single cycle

In the cases of wind and hydro generation the prime mover is connected directly or through a gear box to the alternator. Where there is a heat input the straight system involves:

1 a boiler
2 a superheater
3 a turbine
4 a condenser
5 auxiliary plant such as feed heaters, air ejectors, evaporators and possibly cooling towers.

We will examine these in more detail in conjunction with Figure 1.1.

- **Boiler.** Power station boilers comprise a tall, rectangular open space (like a large luggage lift shaft without the lift with a floor and perhaps four or five storeys high). The walls are of firebrick inside a sheet-steel casing. Inside the firebrick and supported by it all round the walls are vertical steel pipes typically 10 cm in diameter and spaced at 20 cm centres. All the tubes are expanded into one or two large steel drums at the top and smaller ones at the bottom. When operating, the top drums are half full of water so that the tubes and bottom drums are completely full of water.

  The fuel being burnt feeds in through ports near the bottom of the boiler and in burning boils the water and brings it up to the required pressure.
- **Superheater.** Across the top of the boiler there are several layers of spaced tubes through which the combustion products, still very hot, pass. Steam is taken from the top steam drum and fed into these tubes. In passing through it becomes hotter, up to 550 °C being typical.
- **Turbine.** This comprises many sets of blades on its rotor matched by similar sets with opposite pitch on the fixed casing. The steam from the superheater is fed though nozzles on to the first set of rotating blades which creates a force and motion. As the steam exits from this set of blades it is redirected by a fixed set on to the next rotating blades etc. After many such redirections through blades which are longer at each stage the steam approaches the condenser.
- **Condenser.** This is a large vessel across whose width a large number of small tubes pass. These are fed with a constant supply of cold water. The steam in striking these tubes turns rapidly into water which since it has only approximately 1/1700 times the volume of the steam creates a near-perfect vacuum. This very strong suction causes the steam to expand much more than if it exhausted into the atmosphere as in a steam railway locomotive and so to do much more work, making the turbine more efficient. In addition, we have regained the water which can now be pumped back into the boiler to create more steam.
- **Auxiliary plant.** Rather than pump cold water directly back into the boiler it is heated using some of the steam passing through the turbine. At various stages in the turbine there are

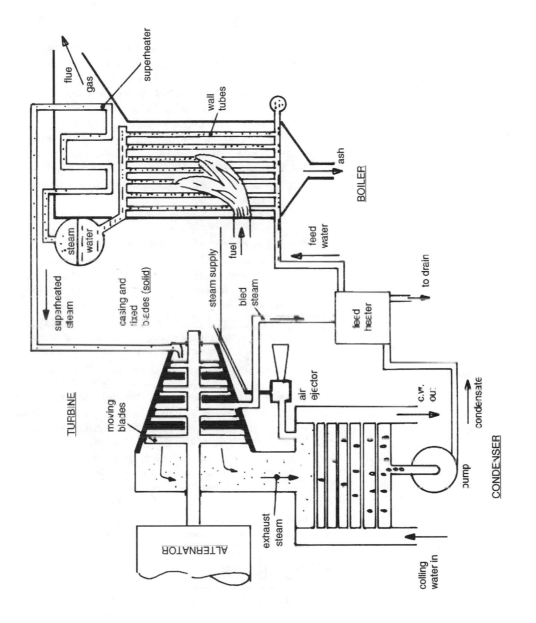

Figure 1.1

points between the fixed blading from which steam can be extracted after doing part of its useful work in driving the shaft. Figure 1.1 shows one such stage but in practice there will be five or more separate steam tap-off points, each one at a higher pressure and hence greater temperature than the last. The steam is taken to tubed water heaters, the feed water to the boiler passing over the steam heated tubes increasing its temperature in each so that when it enters the boiler it is at close to boiling point already.

The vacuum inside the condenser would gradually run down due to air leakage into the casing through the turbine shaft seals so that some of the boiler steam is used to operate an air ejector, constantly sucking air out of the steam space.

There will always be leaks of steam out of the turbine since no seal at the high-pressure end of the shaft is perfect. Gradually there will be less and less water available to pump back into the boiler. Again a small part of the steam passing through the turbine is taken away and used in another heater to evaporate river or sea water which then condenses into pure water for use in the boiler. This evaporator is not shown in Figure 1.1.

The cold water passing through the condenser becomes heated by perhaps 8 °C. Where river or sea water is used this is a once-through process, the hot water returning to its source. Where water is in short supply it has to be cooled down and reused. This is done in cooling towers. The water is pumped through nozzles situated about halfway up the tall tower. The fine spray falling comes into contact with the air which, in being heated, rises to the top of the tower giving the characteristic plume which comes from a cooling tower. The same plume is apparent in cold weather on the top of tall buildings with air conditioning where excess heat is dumped using this method.

When the flue gas leaves the superheater it is still at a very high temperature. It goes on to pass through air heaters which heat the incoming air which is drawn from the top of the boiler house so that it is already warm. Where pulverized coal is used as the fuel the hot air is used to dry the coal so that it may readily be milled into dust which burns with more hot air in a similar manner to a gas or oil flame. Where gas or oil is used the hot air is fed in with the fuel, so maximizing flame temperature and thus efficiency. Where coal is the fuel the flue gas is finally cleaned using some form of dust removal such as an electrostatic precipitator to minimize dust pollution. The chimney temperature is of the order of 140 °C.

Combined cycle

A more efficient use of energy is possible by either producing electricity and useful heat at the same time or generating electricity in two stages. Where heating and electricity are required this can be accomplished in two ways: (1) Use the same cycle as above but extract more steam from the turbine at various bleed stages to heat water for buildings or an industrial process. (2) Instead of using a condenser, cause the steam to leave the turbine at a pressure suitable for any process in mind. For example, a paper works might want a supply of steam at 12 bar to dry the wet paper. The turbine can be designed to exhaust at this pressure using steam from the boiler at, say, 40 bar, thus we get electricity and heat from the system.

Electricity may be generated in two stages by using a gas turbine followed by a conventional boiler. A gas turbine fired with natural gas or oil produces a very hot exhaust at the jet pipe and this jet from a single engine or two engines may be used to drive a turbine connected to an alternator. Aircraft jet engines are available which are capable of generating over 15 MW each. The gas as it leaves the generating turbine is still hot enough to raise steam in a conventional boiler so that another generating stage as above can be added or the boiler heat used in an industrial process.

A very simple electricity/heat combined cycle could involve simply using a diesel engine to drive an alternator whilst using the cylinder cooling water to heat a building or water for industrial process use etc. The heat from this water is radiated into the atmosphere as waste when the engine is associated with a vehicle.

Economics of generation and transmission

The power in a single-phase circuit $= VI \cos \phi$ watts where V and I are the r.m.s. values of circuit voltage and current respectively and ϕ is the phase angle between the current and voltage.

As an example consider a power of 1 MW at 240 V and a power factor of 0.8 lagging. (1 MW $= 1000$ kW $= 10^6$ watts.)

$$240 \times I \times 0.8 = 10^6$$

$$I = \frac{10^6}{240 \times 0.8} = 5208 \text{ A}$$

By increasing the voltage to, say, 20 000 V the required current falls to 62.5 A. The voltage drop in a transmission line due to the resistance of the line $= IR$ volts.
The power loss $=$ voltage drop multiplied by the current flowing
$$= IR \times I$$
$$= I^2 R \text{ watts}$$
Using the above values of current it may be deduced that:

1 for a conductor of given size and resistance, the line losses at 240 V and 5208 A will be very much greater than at the higher voltage; or
2 if the losses are to be the same in both cases the conductor for use at 240 V will need to have a very much lower resistance and hence have a much greater cross-sectional area than for use at the higher voltage.

Example 1.1

Each core of a two-core cable feeding a load 200 m distant from the supply point has a resistance of 0.05 Ω. The voltage at the load end of the cable is 230 V. The power developed in the load is 19.3 kW at 0.85 power factor lagging.
Calculate: (a) the current in the cable; (b) the losses in the cable; (c) the supply voltage; (d) power input to the cable.

(a) Load power = $VI \cos \phi$ watts
Hence $19\,300 = 230 \times I \times 0.85$
Load current (= cable current)

$$I = \frac{19\,300}{230 \times 0.85} = \textbf{98.7 A}$$

(b) Losses in the cable = $I^2 R$ watts
where R = total resistance, go and return
$= 2 \times 0.05$
Losses $= 98.7^2 \times 0.1 = \textbf{974 W}$

(c) Supply voltage = load voltage + volt drops in cable
$= 230 + 2(98.7 \times 0.05)$
$= \textbf{239.9 V}$

(d) Power input to cable = load power + cable losses
$= 19\,300 + 974$
$= \textbf{20\,274 W}$

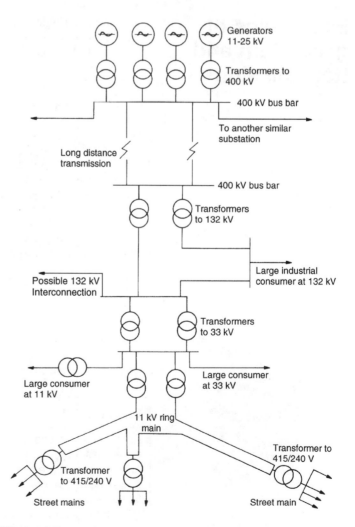

Generators
11-25 kV

Transformers to
400 kV

400 kV bus bar

To another similar
substation

Long distance
transmission

400 kV bus bar

Transformers
to 132 kV

Large industrial
consumer at 132 kV

Possible 132 kV
Interconnection

Transformers
to 33 kV

Large consumer
at 11 kV

Large consumer
at 33 kV

11 kV ring
main

Transformer to
415/240 V

Transformer
to 415/240 V

Street mains

Street main

Figure 1.2

Activity 1.1

In the UK domestic premises are supplied at between 230 V and 240 V. In some other countries 110 V systems are used. Investigate the implications of this in terms of conductor size for specific powers, e.g. for a 100 W light bulb, for a 1 kW fire and for a larger power necessary for an electric cooker. Are there any other considerations, such as safety, sizes of sockets, effects of contact resistances in sockets and switches?

From Example 1.1 and Activity 1.1 it will be apparent that to enable large powers to be transmitted through small conductors whilst keeping the losses small requires the use of high transmission voltages. The voltages commonly in use in the UK are shown in Figure 1.2. At each stage circuit breakers would be used but these are not shown for simplicity.

Generators in major power stations feed directly into step-up transformers which increase the voltage to 400 kV for transmission. Power may be generated locally by, for example, wind turbines or by engines fed with refuse-site methane gas at lower voltages and injected into the system at 11 kV, 33 kV or 132 kV. Power is transmitted to the major load centres where transformers provide the necessary voltages; 132 kV for large industrial consumers and for local distribution to 33 kV substations which in turn feed local industry and a series of 11 kV substations. Ring mains at 11 kV feed transformers which supply power to domestic and commercial consumers at 415/240 V.

To save money on transformers it would seem that generation could best be carried out at 400 kV but so far it has not been found possible to develop insulation for use in rotating machines which will withstand such high voltages while allowing the heat produced in the winding to be dissipated. In particular these are problems where the conductors leave the slots in the iron core and emerge into the gas-filled spaces at the ends of the stator. The problems are overcome in transformers for use at 400 kV by the use of paper insulation and immersing the windings and core completely in a special oil which insulates electrically and convects heat away.

Cables for extra-high-voltage work are also paper insulated and contain oil under pressure. They are laid individually and heat is conducted away by the soil.

As the voltage is increased, as we have already seen, the size and hence the cost of the conductors decreases. However, as the voltage is increased the cost of the insulation is increased. Cable insulation becomes thicker, oil is used and this must often be maintained under pressure which requires additional plant. Very expensive cable terminations called *sealing ends* have to be used.

Switchgear for use at high voltages is more complicated, bulkier and more expensive than that for use at medium and low voltages. When a circuit breaker opens to interrupt a circuit an arc is drawn between the contacts. At domestic voltages the arc is small and arc extinction occurs quickly in the atmosphere. At extra-high voltages the arc is much more difficult to extinguish and air or oil, often under pressure, has to be used to blow out the arc. In addition all electrical parts must be kept well away from earth and these clearances are much greater where very high voltages are used.

The capital costs of extra-high-voltage gear reflect voltage levels but are not affected very much by the cross-sectional area of the conductors used.

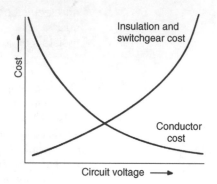

Figure 1.3

Figure 1.3 shows comparative cost of conductors and insulation for increasing system voltages. In addition to the cost of equipment there is the provision of land to consider. The additional bulk of extra-high-voltage gear means that substations may occupy land areas of hundreds or even thousands of square metres.

In Figure 1.2 we see that major transmission is at 400 kV. Since the power transmitted from a single major power station may be in excess of 2000 MW, the high cost of insulation and switchgear is justified by the considerable reduction in conductor costs. At distribution level the supply for a single factory or for housing from an individual transformer represents a relatively small power so that even at much lower voltages the current involved is quite small. The cost of extra-high-voltage switchgear would not be justified and the land area for a substation might well be restricted.

As local demand decreases the voltage at which they are supplied is reduced. A large factory requiring 100 MW will be fed directly from the 132 kV or 33 kV system. A smaller factory requiring only 1 MW could be fed from the 11 kV system whilst a group of houses and shops with a collective requirement of 500 kW will be fed at 415/240 V. The conductor cross-sectional areas generally lie between 225 mm^2 and 650 mm^2 irrespective of voltage, the insulation and switchgear costs and the land area per substation decreasing at each successive voltage reduction.

Typical transformer ratings at the various voltage levels are:

25.6/400 kV	600 MVA
400/132 kV	150–250 MVA
132/33 kV	50–75 MVA
33/11 kV	10–15 MVA
11 kV/415/240 V	250–500 kVA

Overhead lines

Overhead lines for power transmission are almost invariably made of aluminium with a steel core for strength. The bare conductors are supported on insulators made of porcelain or glass which are fixed to wooden poles or steel lattice towers. Figure 1.4 shows some typical British line supports together with the associated insulators. All the steel lattice towers use suspension insulators whilst the wooden poles may use either type. Three conductors comprise a single circuit of a three-phase system so that the 33 kV single circuit towers has three cross arms and three suspension insulators. Towers with six cross arms carry two separate circuits.

On high-voltage lines each support must carry a consecutive recognition number and a circuit-identifying colour disc. The supports must be capable of supporting the line without movement in the ground when both line and supports are carrying a specified ice loading and an 80 km/hr wind is blowing. Safety factors of 2.5 for steel towers and 3.5 for wood must be allowed. A safety factor of 2.5 means that with ice and wind loading the load is 1/2.5 of that which would cause the support to collapse.

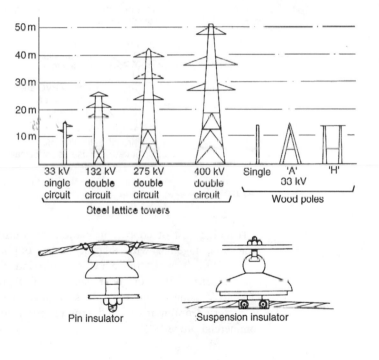

Figure 1.4

Comparison between overhead lines and underground cables

Cost

The overhead line is air insulated and is supported on insulators mounted on towers or poles which are 100–400 m apart. The underground cable is fully insulated and armoured to protect it against mechanical damage and then covered overall with a corrosion-resistant material.

For extra-high-voltage work the overhead line is made of steel-cored aluminium whilst an underground cable is made of copper to reduce the resistance of a given cross-sectional-area cable.

Local underground distribution cables may use solid aluminium cores which are insulated with PVC. Four such cores are laid up, armoured and served overall to form a three-phase cable with the fourth conductor as neutral or earth connection. This is shown in Figure 1.5 together with an oil-pressure cable, three of which are required to form a three-phase circuit.

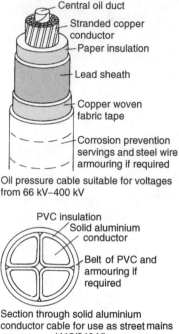

Central oil duct

Stranded copper conductor

Paper insulation

Lead sheath

Copper woven fabric tape

Corrosion prevention servings and steel wire armouring if required

Oil pressure cable suitable for voltages from 66 kV–400 kV

PVC insulation
Solid aluminium conductor

Belt of PVC and armouring if required

Section through solid aluminium conductor cable for use as street mains (415/240 V)

Figure 1.5

The high cost of copper, insulation, armouring and corrosion protection, together with that of taking out a suitable trench and refilling it, make the e.h.v. underground cable many times more expensive than the overhead line. The price difference at 415/240 V using aluminium cables is quite small and these may be preferred on environmental grounds for feeding housing and commercial property.

Environment

The underground cable is invisible. However, there can be no building over it or large trees planted since in the event of a fault it must be possible to dig a suitable hole to effect a repair. The heat produced by an e.h.v. cable can affect the soil around it thus modifying the plant growth in the immediate vicinity.

The overhead line has conductors and supports which are sometimes visible for long distances. Electrical discharges from the lines can cause radio interference. A possible problem which has been recently highlighted is the effect that electro-static and electro-magnetic fields produced by high-voltage lines may have on people living in the vicinity of these lines.

Reliability

There is little difference between the reliabilities of the two systems. The overhead line can be struck by lightning whereas the underground cable is at the mercy of earth-moving machinery, especially when roads are remade or trenches for other services are dug. Occasionally a cable will develop a small hole due to movement over a stone, for example, giving rise to water ingress followed by an explosion, but this is thankfully rare.

Fault finding

Overhead lines are patrolled regularly on foot or by helicopter. Broken insulators can be seen and by using infra-red detection equipment local hot spots can be found, possibly in compression joints where two lengths of conductor have been joined. Repairs are reasonably cheap since the line can be taken down, insulators replaced, joints remade and the towers repainted at almost any time. If an underground cable develops a fault electrical methods have to be used to locate it. Unless the route is precisely known and the test accurately carried out a great deal of digging is required before the fault can be found. When it has been located the repair is expensive, especially on e.h.v. cables.

Activity 1.2

You are a leader of a team of engineers engaged to advise on energy matters. You have been retained by a fairly large firm who purchase and machine wood; this activity produces, amongst other things, large volumes of wood chippings. One method of disposal would be to burn these to produce (1) heat for the processes used in the factory and (2) electricity for the factory and administrative buildings, selling the excess to the local supply authority.

The factory management are not aware of the costs involved in the National supply of power. Write a report to them explaining why it is that although major power stations can generate power at a lower cost per kWh than you can, your cost can still be viable. (Explain how the costs of power bought from the area supplier are made up, and how savings can be made by generation at lower voltages by avoiding the losses involved with transformers and high-voltage transmission systems.) Explain how industrial heat and electricity can be produced simultaneously so that the electricity produced is almost a by-product. (Investigate *pass out* and *bled steam* industrial turbines.)

Switchgear, definitions and uses

- **Circuit breaker.** A circuit breaker is a mechanical device for making and breaking a circuit under *all* conditions.
- **Switch.** A switch is a device for making and breaking a circuit which is carrying current not greatly in excess of normal loading.
- **Isolator.** An isolator is a means of isolating or making dead a circuit which is not carrying current at the time (like pulling out a fuse at home so that work may be carried out safely on a circuit). The isolator may be used to close the circuit on to load.

The Electricity Supply Regulations state that no piece of electrical equipment may be connected to the mains unless the circuit incorporates a device which will disconnect the equipment automatically in the event of a fault.

According to the definitions above, a circuit breaker is such a device. These are made in miniature form for domestic use with current ratings between 5 A and 60 A at 240 V whilst there are larger sizes for industry and transmission and distribution substations which can deal with the highest voltages and currents presently in use.

A fuse is often used in place of a circuit breaker in circuits operating up to 11 kV but once this has operated to clear a fault it has to be replaced. This takes time and the larger sizes are very expensive. A circuit breaker can be reclosed after clearing a fault and in addition it may be used in the role of a switch, making and breaking circuits under normal conditions.

Switchgear may be of either the indoor or outdoor variety. For use indoors all electrical conductors are completely enclosed. For use outdoors the circuit breakers are made completely weatherproof. The circuit conductors and isolators are enclosed in 11 kV to 415/240 V substations but at higher voltages they are bare metal insulated from earth using porcelain or glass insulators as described for overhead lines. Figure 1.6 shows a bulk oil circuit breaker for use outdoors. It is suitable for use up to 132 kV. Figure 1.7 shows an air circuit breaker employed up to about

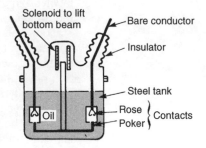

Figure 1.6

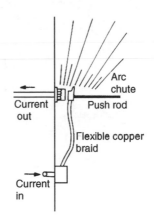

Figure 1.7

11 kV. It may be used indoors and outdoors according to the type of enclosure.

Substation layout and equipment

The small substation shown in Figure 1.8 is typical of the thousands of such installations feeding factories, housing and commercial premises. It is connected to the 11 kV ring main shown in Figure 1.2.

Considering Figure 1.8, the ring main is equipped with two isolators at each tap-off point and these are normally closed making the 11 kV ring complete. The transformer is fed through a circuit breaker. If a fault occurs on the transformer itself the circuit breaker should open. If a fault occurs on the low-voltage system it should be cleared by the low-voltage circuit breaker or one of the fuses. Should a fault occur and the correct clearance not take place the circuit breakers controlling the 11 kV ring (not shown) would operate making the whole ring dead, depriving other consumers of their supply.

Circuit breakers may be of the oil-immersed or air-break type. Switches could be used to control individual low-voltage circuits for lighting and heating and an automatic switch called a contactor would be used for motor control.

The transformer at the substation will be rated at 11 000 to 433 V nominally. On the high-voltage side of the transformer there will be a tap changer. This is a manually operated device which is adjusted off load to alter the number of turns on the winding and has the effect of altering the output voltage to the desired level. The transformer rating will be 300, 500, 750 or 1000 kVA. It will generally be oil-filled using special transformer mineral oil and the heat produced will be carried by the oil to cooling tubes on the outside of the transformer whence it is convected away by natural air circulation. For this reason the building must be suitably ventilated. This is achieved by fitting louvres in the door of the substation and air bricks or more louvres high up in the walls. Very large transformers are fan cooled but these are installed outdoors. The transformer is either mounted on a plinth in the centre of a large pit filled with stones or there is a small wall built around

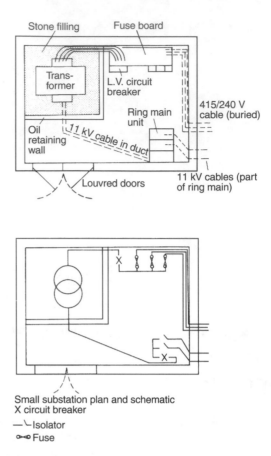

Figure 1.8

it. If the transformer suffers a fault and oil leaks out this must be contained to minimize the risk of fire spreading.

In a very large substation with many circuit breakers fire-resistant barriers are used to create small sections and automatic fire-fighting equipment is fitted in each section using either carbon dioxide gas or fine water spray. Should a fire occur on one circuit breaker, only the equipment in the affected section operates.

Activity 1.3

Organize a visit to a power station and/or substation. Seek out any available literature, question your guide and use the information to enable you to write a detailed report of what you have seen together with block diagrams marked up with their function and importance in the system. Many stations have advertised open days but in any case will usually respond to a letter to the Station Manager. If you are unable to make such a visit, do a library project with the same ends in mind. There is really no substitute for looking at and hearing all that is going on in a real station.

The three-phase system

Star connection

In the three-phase generator the magnetic field on its rotor links in sequence with three equally spaced windings or phases on its stator inducing sinusoidal voltages with equal maximum values in each of them.

These are known as the Red, Yellow and Blue voltages respectively. These are shown in Figures 1.9 and 1.10. The phasors rotate anticlockwise and the red phase voltage is followed by that of the yellow phase and again by that of the blue phase. This is known as a *positive sequence* and is the normal situation. Where the opposite sequence is involved, that is, red followed by blue followed by yellow, this is known as the *negative sequence*. We shall only consider the normal or positive sequence. Figure 1.11 shows a schematic diagram of a generator, its windings physically displaced from each other by 120° and with one end of each winding connected to earth. Because of its appearance this is known as the star connection and the common point for the three windings is called the *star point*. The supply lines are labelled red, yellow and blue respectively and the wire connected to the star point is known as the neutral. The voltage from any output line to the neutral is called the phase voltage and the voltage between any pair of supply lines is called the line voltage.

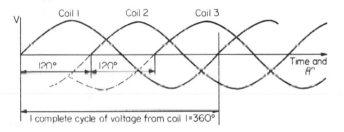

Figure 1.9

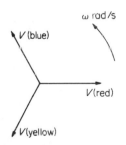

Figure 1.10

Activity 1.4

Measure the voltages on a three-phase system
(1) between each of the lines and the neutral; (2) between each of the lines, red to yellow, yellow to blue and blue to red.

Are all the readings for (1) identical and for (2) identical? If they are slightly different in each case, can you suggest a reason?

Mathematically divide the magnitude of the line voltage by that of the phase voltage.

Draw your phase voltages to scale as in Figure 1.10. Scale the distance between the phasor tips. What is the relationship between phase and line voltages? Is this relationship borne out by the phasor diagram?

Using a multi-range volt meter see if there is a potential difference between the neutral wire and earth. (You may find 0.5–2 V here.) Why is this, since the transformer neutral is connected to earth at the transformer end?

From Activity 1.4 you should have determined that line voltage

$$= \sqrt{3} \times \text{phase voltage}$$

$$V_L = \sqrt{3} V_{PH}$$

For example, where the phase voltage is 240 V, the line voltage is $\sqrt{3} \times 240 = 415$ V. Single-phase loads are connected between any line and the neutral wire and this is the normal situation in the home. One house in the street is connected between the red line and neutral, the next house between yellow line and neutral and the next from blue line to neutral.

In industry three-phase loads may use three or four wires. Some loads are delta-connected (from the Greek letter Δ [delta]) when each of the load components is connected directly from line to line, red to yellow, yellow to blue and blue to red. This means that each of the components is subjected to line voltage which is $\sqrt{3} \times$ phase voltage and more power will be dissipated. However, generation and distribution are always carried out using the star connection because this provides a neutral point which is connected to earth providing the means to protect the circuit in the event of an earth fault on one of the lines. In the home an earth fault

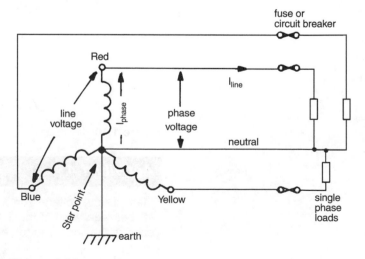

Figure 1.11

on a piece of equipment causes current to flow down the earth wire and to reappear at the supply transformer earthed star point. This current should be large enough to clear either the fuse or miniature circuit breaker protecting the circuit. Protection can be made more sensitive by using a residual current device which compares the currents in the line and neutral wires. If the currents are identical the breaker remains closed. If a current leaks to earth from either wire, the live and neutral currents will no longer balance and the device trips. These devices generally operate with a current difference of 30 mA or 60 mA. An earth fault on the street main will cause sufficient current to flow back to the supply transformer star point to either clear the main fuse or open the circuit breaker.

Generators are protected in the same way. However, since the current flowing in the event of a fault could be extremely large they often have a resistor built into their earth connection to restrict fault current. A current transformer is mounted on the earth connection and in the event of an earth fault a quite small current being detected in the current transformer will cause the generator to trip; the circuit breaker will open and the main steam or fuel valve will be shut off.

Load balancing. Consider now the case of three single-phase loads being equal in magnitude and phase. Figure 1.12(a) shows the three-phase voltages as in Figure 1.10 together with the current phasors. These are equal in magnitude and lag on the phase voltages by an angle $\phi°$. All these currents flow from their respective lines to the neutral. The current in the neutral must therefore be the sum of these three currents.

Figure 1.12(b) shows the phasor addition using two methods. In the first I_R is added to I_B and this is found to be exactly equal in magnitude and in the opposite direction to I_Y. Therefore adding $I_R + I_B$ to I_Y gives zero and the neutral current proves to be zero. In the second method, the phasors are joined together in sequence forming a closed loop, there being no gap or resultant current. This is a recognized way of adding phasors with which the student should be familiar from 'principles' lessons. There is no current flowing in the neutral wire back to the supply generator when the loads are balanced. Considering Example 1.1 in which we considered the voltage drop and power loss in both go and return wires, with the three-phase system, if it is balanced, we need only consider the go or line volt drop. In many three-phase circuits, such as those supplying motors where balance is assured, only three wires are run, the neutral being dispensed with.

It is desirable that over the country as a whole the loads shall be balanced over the three phases since this minimizes the cable losses and voltage drops in the supply lines. In addition, there are problems in the generators themselves when the three phases are not equally loaded.

Let us consider the power loss and voltage drops in an unbalanced system.

The worst out of balance is when all the load is on a single phase. Figure 1.13 shows a resistive load carrying a current of 21 A from a 240 V supply along a cable in which line and neutral resistances are both 0.1 Ω.

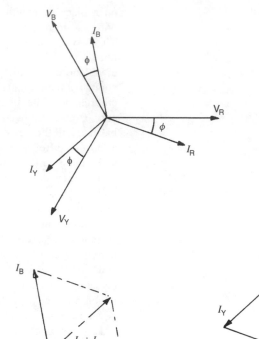

(a)

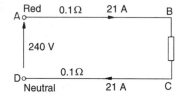

(b)   Two methods of phasor addition

Figure 1.12

Figure 1.13

As in Example 1.1 Power loss $= 21^2 \times 0.2 = 88.2$ W

Load potential difference $= 240 - (21 \times 0.2) = 235.8$ V

Now consider the effect of obtaining the same power (very closely) by using three separate single-phase resistive loads each drawing 7 A as shown in Figure 1.14. There is no neutral current because the loads are balanced.

Power loss in each line $= I^2R = 7^2 \times 0.1 = 4.9$ W

Power loss in three lines $= 3 \times 4.9 = 14.7$ W

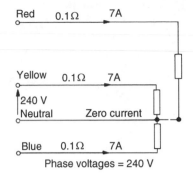

Figure 1.14

Fall of potential along each line $= IR = 7 \times 0.1 = 0.7$ V

Potential difference across each load $= 240 - 0.7 = 239.3$ V

Balancing the loads has eliminated the loss in the return (neutral) wire, reduced the losses and increased the load voltages.

Where the loads are not balanced across the phases there will be a current flowing in the neutral. This current causes the star point of the load to be at a potential above that of earth, this potential being required to drive the current through the neutral wire back to the supply transformer star point. This will involve additional losses as previously discussed.

Example 1.2

A three-phase star-connected system has the following loads connected between the respective phases and the neutral:

> 5 A at unity power factor in the red phase

> 2.5 A at 0.8 power factor leading in the yellow phase

> 4 A at 0.707 power factor lagging in the blue phase.

Draw a phasor diagram showing the phase voltages and currents and determine the value of the neutral current under these conditions.

Unity power factor means that the red phase current is in phase with the red phase voltage.
0.8 power factor leading means that the yellow phase current leads the yellow phase voltage by an angle arc $\cos 0.8 = 36.87°$.
0.707 power factor lagging means that the blue phase current lags the blue phase voltage by an angle arc $\cos 0.707 = 45°$.

Figure 1.15 shows the phase voltages and currents. The additional angles shown are derived by remembering that there are 120° between each of the voltage phasors.

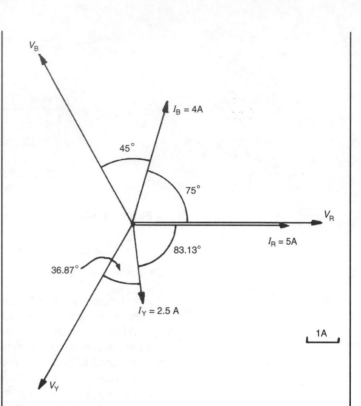

Figure 1.15

The currents are added using a similar method to that shown in Figure 1.12(b) or by resolution into horizontal and vertical components. Figure 1.16 shows the phasor addition. Resolving into horizontal (H) and vertical (V) components yields the following result:

$$I_R = 5 \text{ A (H)} + 0 \text{ (V)}$$

$$I_Y = 2.5 \cos 83.13° \text{ (H)} + 2.5 \sin 83.13° \text{ (V)}$$

$$= 0.3 \text{ A (H)} - 2.48 \text{ A (V)}$$

(−ve sign indicating downwards from the origin)

$$I_B = 4 \cos 75° \text{ (H)} + 4 \sin 75° \text{ (V)}$$

$$= 1.03 \text{ A (H)} + 2.482 \text{ A (V)}$$

Adding horizontal components:

$$(5 + 0.3 + 1.03) = 6.33 \text{ A (H)}$$

Adding vertical components:

$$(0 - 2.48 + 3.86) = 1.38 \text{ A (V)}$$

Resultant neutral current by Pythagoras:

$$I_N = \sqrt{(6.33^2 + 1.38^2)} = \textbf{6.48 A}$$

The resultant neutral current makes an angle with the horizontal (red phase voltage) of arctan $1.38/6.33 = 12.3°$.

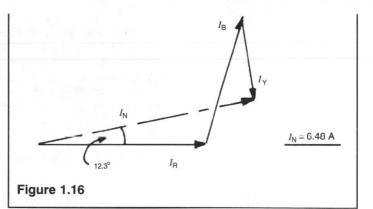

Figure 1.16

The delta connection

Figure 1.17 shows the delta connection. There is no neutral wire. The three-phase loads shown are connected across the lines and are each subjected to line voltage. However, in this case there is a current division to be considered. Applying Kirchhoff's current law to junction X in Figure 1.17 we see that $I_R + I_L - I_Y = 0$.

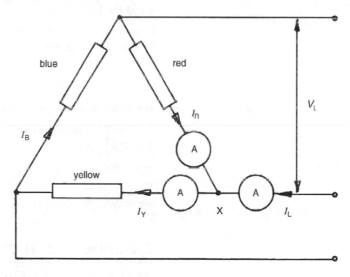

Figure 1.17

Transposing: $I_L = (I_Y - I_R)$.

With the star connection, phase and line currents were the same but phase and line voltages differed. With the delta connection, phase and line voltages are the same but phase and line currents differ.

Activity 1.5

Set up a circuit as shown in Figure 1.17 using three identical resistors and three ammeters. Connect it to a three-phase supply and measure the two phase currents and the line current. Divide mathematically the magnitude of the line current by the common value of phase current. What relationship between the two is there? Reconnect the circuit using another line with the associated two phases. Is the relationship true for this line?

From Activity 1.5 it should be established that $I_L = \sqrt{3}I_{PH}$.

Power in the three-phase balanced load

1 Connected in star

Looking at Figure 1.11 we see a generator supplying a balanced load. Phase voltage for generator and load, ignoring volt drops, $= V_{PH}$.

Considering one phase of the generator, the phase current I_{PH} flows in the supply line and in one of the load phases. The line current is called I_L and for the star connection phase and line currents are equal since, by inspection, current leaving the generator phase has nowhere else to go except into the line and on to the load.

Generally, power in an AC circuit $= VI\cos\phi$ watts
Using the voltage and current defined above, power in one phase $= V_{PH}I_{PH}\cos\phi$ watts

Since $I_{PH} = I_L$ we may write this as $V_{PH}I_L\cos\phi$ watts

For three identical phases, power $= 3 \times V_{PH}I_L\cos\phi$ watts

3 may be expressed as the product of its two square roots, $3 = \sqrt{3} \times \sqrt{3}$

Total power $= \sqrt{3} \times (\sqrt{3}V_{PH})I_L\cos\phi$ watts and since $V_L = \sqrt{3}V_{PH}$
we may write: Total power $= \sqrt{3}V_LI_L\cos\phi$ watts.

2 Connected in delta

Power per phase $= V_{PH}I_{PH}\cos\phi$ watts
In three phases $= 3V_{PH}I_{PH}\cos\phi$ watts

But $V_{PH} = V_L$ and $I_{PH} = I_L/\sqrt{3}$ for the delta connection (established in Activity 1.5)

$$\text{Total power} = 3V_L\frac{I_L}{\sqrt{3}}\cos\phi$$

$$= \sqrt{3}V_LI_L\cos\phi \text{ watts}$$

as for the star-connected system.

Note: when the voltage of a three-phase system is quoted it is always given as a **line** value. A 415 V, three-phase supply has a line voltage of 415 V and a phase voltage of 240 V.

Example 1.3

An 11 kV, three-phase, star-connected alternator delivers a load of 60 MW at 0.85 power factor lagging to its step-up transformer which has a delta-connected primary. It delivers power from its star-connected secondary at 132 kV (line).

Determine (a) the generated phase voltage,
(b) the generator phase current,
(c) the generator line output current,
(d) the phase current in the transformer primary,
(e) the phase and line current in the transformer secondary, and
(f) the phase voltage of the transformer secondary.

(a) $V_L = \sqrt{3} \times V_{PH}$ Therefore $V_{PH} = V_L/\sqrt{3}$
$$= 11\,000/\sqrt{3}$$
$$= \textbf{6350 V}$$

(b),(c) Power $= \sqrt{3}V_L I_L \cos\phi$ watts $(= 60 \times 10^6)$
$$= \sqrt{3} \times 11\,000 \times I_L \times 0.85$$
$$I_L = \frac{60 \times 10^6}{\sqrt{3} \times 11\,000 \times 0.85} = \textbf{3705 A}$$

and with the star connection $I_L = I_{PH}$

(d) The phase current in the delta-connected primary
$$= I_L/\sqrt{3}$$
$$= 3705/\sqrt{3}$$
$$= \textbf{2138 A}$$

(e) The transformer secondary operates at a line voltage of 132 000 V
$$I_L = \frac{60 \times 10^6}{\sqrt{3} \times 132\,000 \times 0.85} = \textbf{308.7 A}$$

and with the star connection $I_L = I_{PH}$

(f) Phase voltage in star $= V_L/\sqrt{3} = 132\,000\sqrt{3}$
$$= \textbf{76\,200 V}$$

The quantities are illustrated in Figure 1.18.

Test your knowledge 1.2

A three-phase, delta/star transformer is fed from an 11 kV supply and delivers 500 kW at 0.8 power factor lagging to a small factory at a line voltage of 415 V. Ignoring transformer losses and assuming perfect balance, determine the values of (a) the input line current, (b) the primary phase current, (c) the line current delivered to the factory.

Industrial installations

A typical substation for a factory is shown in Figure 1.8. Such a substation would most likely be situated on the factory premises.

Metering of energy consumed and the maximum demand made by the factory on the supply system is carried out using current transformers fitted in, or adjacent to, the main low-voltage circuit breaker or fuses. The current transformers feed a kilowatt-hour meter which at 415/240 V derives the supply for its voltage coil directly from the bus bars without the use of potential transformers. The main distribution fuse board is fitted in the substation.

In Figure 1.19 this is shown as being equipped with switches incorporating fuses rated at 400 A. The actual size employed will depend on the rating of the equipment used in the factory.

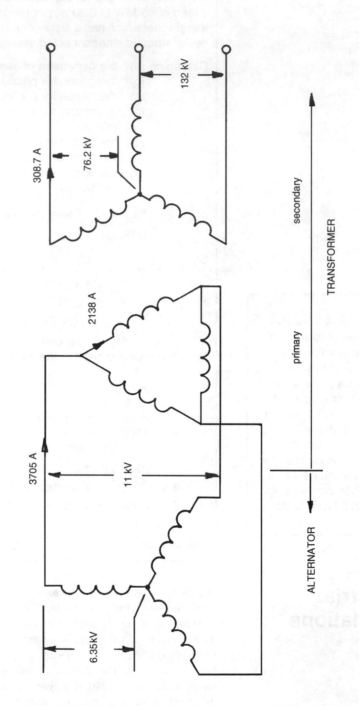

Figure 1.18

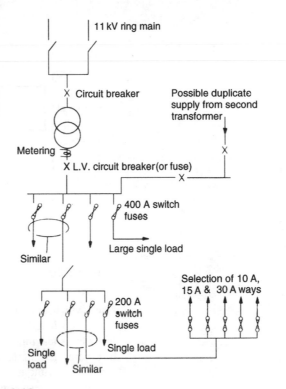

Figure 1.19

Some large single loads may be fed directly from this point, the actual control being by local switch or contactor. There will be several ways feeding further fuse boards in particular workshops. These will feed heavy local loads or further fuse boards for lighting, heating, small tools and processes.

In Figure 1.19 single-line representation has been used: instead of drawing three conductors representing the three phases throughout the diagram, a single line and a single fuse are used to represent a three-phase system. This means, for example, that a single 400 A switch contains in fact three fuses and when it operates, all three phases are broken. The fuse board shown, with 10 A, 15 A and 30 A ways, will have these spread, as far as possible, over the three phases.

On single-phase systems, although there will be two wires for each circuit, often only one line is drawn representing the live conductor in a single cable; the neutrals are all common and not shown. In tables of resistance and volt drops these will be quoted as a 'loop' value, that is, the values for both go and return conductors are lumped together.

Distribution systems

Radial system

A radial system for the distribution of electrical energy is shown in Figure 1.20. A substation supplies consumers C through radial distributors which fan out from the substation. A fuse or circuit breaker protects each distributor. Some distributors have subsidiary

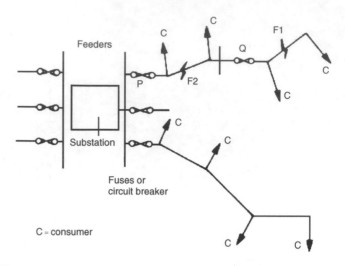

Figure 1.20

fuses which are of lower rating than the main fuse somewhere along their length. In the event of a fault on the feeder the relevant fuse clears leaving all consumers in that section without a supply. In Figure 1.20 a fault F1 would cause fuse Q to clear leaving two consumers without a supply. Since there is no alternative method of supply to these consumers repairs to the line have to be carried out before they can be reconnected.

The consumer on the end of each distributor suffers voltage reductions as the load on that distributor increases and to minimize this the cross-sectional area of the conductors is large. This is therefore an expensive system to install and offers poor security of supply.

Fault finding is relatively simple since tests need only be done on the cleared section of line.

Ring system

Addition of further substations feeding the other ends of the radial feeders as shown in Figure 1.21 effectively converts the system into a ring.

This substantially reduces the voltage drops along the distributors and enables savings in conductor cross-sectional areas and costs to be made. A fault F1 would result in fuses Q and R clearing leaving two consumers without a supply. A fault F2 would result in fuses P and Q clearing and again only two consumers would lose their supply instead of four with the radial system.

At 11 kV a ring system employing isolators at each load point enables greater security of supply to be achieved. In Figure 1.22 a fault F1 can be cleared by opening isolators I_2 and I_3 and no other section need be disconnected provided that a second fault does not occur. Meanwhile repairs can be carried out.

Interconnection of two points on the ring makes the system more versatile while reducing voltage drops and cable losses. This makes for greater expense, however.

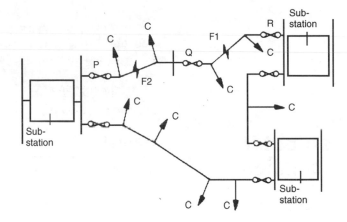

Figure 1.21

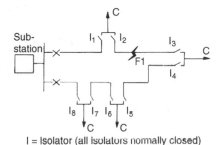

I = Isolator (all isolators normally closed)

Figure 1.22

If circuit breakers are used instead of isolators, overcurrent relays which are sensitive to the direction of current flow can be used to automatically isolate a faulty section. When isolators are used there will be an interruption of supply to all consumers on the ring as the main breakers open. After locating the fault the relevant isolators are manually operated when the main breakers can be closed, restoring supply.

Reduction of voltage drops and cable losses together with increased security of supply is achieved by increasing the number of substations and the degree of interconnection. However, the expense rapidly increases with complexity and the actual arrangement employed is the best that can be obtained at realistic cost to the consumer.

Activity 1.6

Make a distribution diagram showing supply points, main control gear and subcircuits for an industrial or commercial property. The data could be obtained by visiting a factory or perhaps a supermarket and making notes as to where the supply enters the building; at what voltage; what are the control mechanisms, where are the fuse boards, what are the fuse ratings etc. What type of cabling is used for the supply and distribution? Involve your local electricity

supply company, perhaps arranging a visit to their local
control room. Try to discover the basis for their tariffs; are
they based on maximum demand or do they have different
rates for different times of the day? Write a report on your
findings to support your system diagrams.

Distributor calculations

Radial

In Figure 1.13 we saw a single resistive load being fed from a
240 V supply. Using single-line representation the diagram can be
redrawn as shown in Figure 1.23.

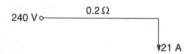

Figure 1.23

The resistance marked is that of both go and return conductors.

Load voltage $= 240 - (21 \times 0.2)$

$\qquad = 240 - 4.2$

$\qquad = 235.8$ V as previously.

The method is suitable for both DC and single-phase AC where
resistance only is considered. When the loads or lines have
inductance or capacitance, there are phase angles to consider and
arithmetic addition and subtraction of voltages gives incorrect
results.

Let us consider a feeder with three loads.

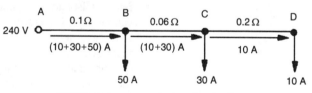

Resistances are go and return (loop) values

Figure 1.24

Example 1.4

The details of a radial feeder are shown in Figure 1.24.

Calculate (a)   The load voltages
         (b)   The power lost in the cable
         (c)   The power developed by each load
         (d)   The efficiency of the system.

Applying Kirchhoff's first law to each load point:

The current in section CD of the feeder = 10 A

Section BC carries load currents C and D $= 30 + 10$
$$= 40 \text{ A}$$

Section AB carries the total load currents $= 40 + 50$
$$= 90 \text{ A}$$

The voltage drop between A and B $= I_{AB} \times R_{AB} = 90 \times 0.1$
$$= 9 \text{ V}$$

Voltage at load B $= 240 - 9 = \textbf{231 V}$

Power loss in section AB $= (I_{AB})^2 R_{AB} = 90^2 \times 0.1$
$$= 810 \text{ W}$$

Power developed by load B $= V_B I_B = 231 \times 50$
$$= 11\,550 \text{ W}$$

Repeat for section BC

Volt drop from B to C $= 40 \times 0.06 = 2.4$ V

Voltage at load C $= 231 - 2.4 = \textbf{228.6 V}$

Power loss in section BC $= 40^2 \times 0.06 = 96$ W

Power developed by load C $= 228.6 \times 30 = 6858$ W

Repeat for section CD

Voltage drop from C to D $= 10 \times 0.2 = 2$ V

Voltage at load D $= 228.6 - 2 = \textbf{226.6 V}$

Power loss in section CD $= 10^2 \times 0.2 = 20$ W

Power in load D $= 226.6 \times 10 = 2266$ W

Total load powers $= 2266 + 6858 + 11\,550 - \textbf{20\,674}$

Total losses $= 20 + 96 + 810 = \textbf{926 W}$

$$\text{Efficiency} = \frac{\text{Power in loads}}{\text{Total power input}}$$

$$= \frac{20\,674}{20\,674 + 926} \quad \text{OR} \quad \frac{20\,674}{240 \times 90}$$

$$= \textbf{0.957 p.u.} \text{ or } \textbf{95.7\%}.$$

Distributor fed at both ends

Now consider the effect of reinforcing the system by feeding point D either from another substation or from the same substation through an extra length of cable or line.

Example 1.5

Recalculate (a) to (d) in Example 1.4 with the supply reinforced as shown in Figure 1.25.

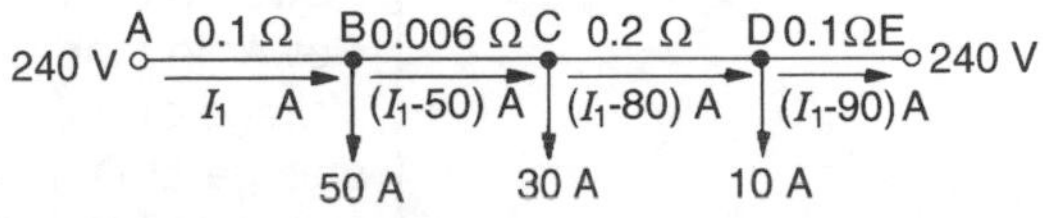

240 V $\circ$ —— A 0.1 Ω —— B 0.006 Ω C 0.2 Ω —— D 0.1 Ω E $\circ$ 240 V

I_1 A (I_1-50) A (I_1-80) A (I_1-90) A

50 A   30 A   10 A

Figure 1.25

In this case we have first to determine how much current is supplied from each end of the feeder.

Consider a current I_1 to be entering from the left-hand end. The current in this section $AB = I_1$ A and the volt drop between A and $B = I_1 \times 0.1$ V.

At point B a load current of 50 A is supplied so that the current flowing on in section BC must be $(I_1 - 50)$ A and the volt drop from B to $C = (I_1 - 50) \times 0.06$ V.

Similarly the current in section $CD = (I_1 - 50) - 30 = (I_1 - 80)$ A and the volt drop from C to $D = (I_1 - 80) \times 0.2$ V.

In section DE the current is $(I_1 - 90)$ A and the voltage drop from D to $E = (I_1 - 90) \times 0.1$ V.

The voltage at A – all the volt drops along the line = voltage at E.

This is the same as Example 1.2 except that now we know the voltage at E is held at 240 V.

Therefore
$$240 - I_1 \times 0.1 - (I_1 - 50) \times 0.06 - (I_1 - 80) \times 0.2$$
$$-(I_1 - 90) \times 0.1 = 240 \text{ V}$$

Multiply out the brackets
$$240 - 0.1I_1 - 0.06I_1 + 3 - 0.2I_1 + 16 - 0.1I_1 + 9 = 240$$

Transpose
$$240 - 240 + 3 + 16 + 9 = 0.1I_1 + 0.06I_1 + 0.2I_1 + 0.1I_1$$
$$28 = 0.46I_1$$
$$I_1 = 60.86 \text{ A}$$

Redraw the diagram as shown in Figure 1.26.

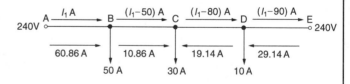

Figure 1.26

In section BC we have $(60.86 - 50) = 10.86$ A

In section CD we have $(60.86 - 80) = 19.14$ A. The negative sign indicates a reversal of current direction from that shown in Figure 1.25 and this is seen to be logical since the 30 A load receives 10.86 A from end A and 19.14 A from end E.

Voltage at B $= 240 - 60.86 \times 0.1 = $ **233.9 V**.

Power loss in section AB $= 60.86^2 \times 0.1 = 370.4$ W.

Power in load B $= 233.9 \times 50 = 11\,695$ W.

Voltage at C $= 233.9 - 10.86 \times 0.06 = $ **233.25 V**.

Power loss in section BC $= 10.86^2 \times 0.06 = 7.08$ W.

Power in load C $= 233.25 \times 30 = 6997.5$ W.

Voltage at point D. Since current flows from D to C the voltage at D must be greater than that at C. The minimum voltage on the distributor is at load C and currents flow from each end to this point.

Test your knowledge 1.3

A two-wire feeder is fed at end A at 240 V and at end B at 250 V. Loads are connected to the feeder as follows:

(a) 10 A at point B which is 100 m from A
(b) 100 A at point C which is 150 m from B
(c) 70 A at point D which is 200 m from C
(d) 25 A at point E which is 50 m from D and 25 m from B.

The resistance of the feeder is 0.08 Ω per 100 m LOOP. Determine:
(a) the minimum potential on the feeder
(b) the efficiency of the system.

Voltage at D $= 233.25 + 19.14 \times 0.2 = $ **237.08 V**.
Power loss in section CD $= 19.14^2 \times 0.2 = 73.26$ W.
Power in load D $= 237.08 \times 10 = 2370.8$ W.
We know that the voltage at point E is 240 V. Let us check the calculations by adding the volt drop from E to D to the voltage at D.
$237.08 + 29.14 \times 0.1 = 240$ V
Power loss in section DE $= 29.14^2 \times 0.1 = 84.91$ W.
Total load powers $= 11\,695 + 6997.5 + 2370.8$
$$= \textbf{21\,063 W}.$$
Total losses $= 370.4 + 7.08 + 73.26 + 84.91$
$$= \textbf{535.65 W}.$$

$$\text{Efficiency} = \frac{21\,063.3}{21\,063.3 + 535.65} = \textbf{0.975 p.u.}$$

Notice that by feeding the system at both ends each of the load voltages has been increased. The cable losses have been reduced so increasing the efficiency.

Example 1.6

For the ring main shown in Figure 1.27 determine the current in each section and the minimum load voltage.

Redraw the diagram putting in the go and return resistances. (Observe that in the diagram, the resistance of ONE CONDUCTOR is given per 100 m.) A 100 m go and return has a resistance of $2 \times 0.05 = 0.1$ Ω. A ring main may be considered as a feeder fed at both ends at the same voltage.

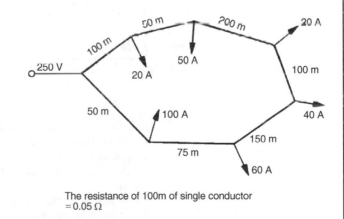

The resistance of 100m of single conductor $= 0.05$ Ω

Figure 1.27

Starting with a current I_1 flowing towards the 20 A load:

$$250 - 0.1I_1 - 0.05(I_1 - 20) - 0.2(I_1 - 70) - 0.1(I_1 - 90)$$
$$- 0.15(I_1 - 130) - 0.075(I_1 - 190)$$
$$- 0.05(I_1 - 290) = 250$$

$$250 - 0.1I_1 - 0.05I_1 + 1 - 0.2I_1 + 14 - 0.1I_1 + 9$$
$$- 0.15I_1 + 19.5 - 0.075I_1 + 14.25$$
$$- 0.05I_1 + 14.5 = 250$$
$$72.25 = 0.725I_1$$
$$I_1 = \textbf{99.65 A}$$

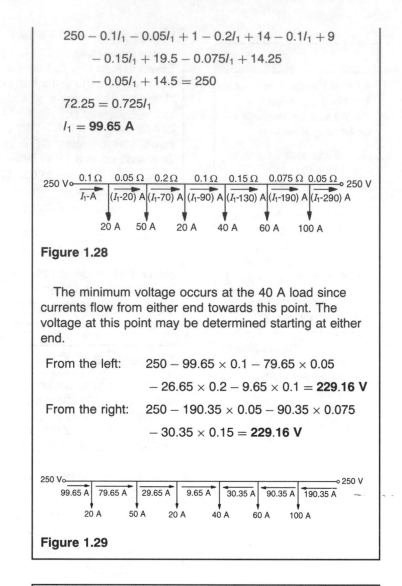

Figure 1.28

The minimum voltage occurs at the 40 A load since currents flow from either end towards this point. The voltage at this point may be determined starting at either end.

From the left: $250 - 99.65 \times 0.1 - 79.65 \times 0.05$
$$- 26.65 \times 0.2 - 9.65 \times 0.1 = \textbf{229.16 V}$$

From the right: $250 - 190.35 \times 0.05 - 90.35 \times 0.075$
$$- 30.35 \times 0.15 = \textbf{229.16 V}$$

Figure 1.29

Example 1.7

A two-wire feeder is 275 m in length. It is fed at 238 V at one end and 245 V at the other end. The feeder loop resistance (go and return conductors) is 0.05 Ω/100 m. Three loads of 50 A, 100 A and 25 A are being fed at distances of 100 m, 150 m and 200 m respectively from the end of the feeder being fed at 238 V.

Calculate (a) the value of the minimum load terminal voltage on the system and (b) the overall efficiency of the system.

First draw the system and, using the distances, calculate and mark on the feeder the loop resistances (Figure 1.30). For example, 75 m of feeder will have a resistance of $75/100 \times 0.05 = 0.0375 \ \Omega$.

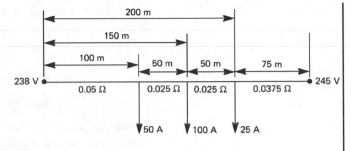

Figure 1.30

Let I_1 A enter at the left-hand end. The currents will be as shown in Figure 1.31.

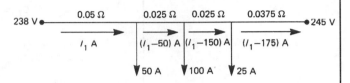

Figure 1.31

$238 -$ (all the voltage drops along the feeder) $=$ voltage at the other end $= 245$ V

$238 - 0.05I_1 - 0.025(I_1 - 50) - 0.025(I_1 - 150)$
$\quad - 0.0375(I_1 - 175) = 245$

$238 - 0.05I_1 - 0.025I_1 + 1.25 - 0.025I_1 + 3.75$
$\quad - 0.0375I_1 + 6.56 = 245$

$238 + 1.25 + 3.75 + 6.56 - 245$
$\quad = (0.05 + 0.025 + 0.025 + 0.0375)I_1$

$4.56 = 0.1375I_1$

$I_1 = 33.2$ A

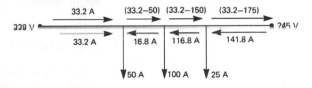

Figure 1.32

The minimum terminal voltage is at the 50 A load, since currents flow towards this load from both ends (Figure 1.32).

(a)   Load voltage $= 238 - 33.2 \times 0.05$

$$= \mathbf{236.34 \ V}$$

The same result is achieved by working from the other end of the feeder, subtracting the voltage drops from 245 V.

(b)   Losses $= 33.2^2 \times 0.05 = 55.1$ W

$16.8^2 \times 0.025 = 7.06$ W

$116.8^2 \times 0.025 = 341.06$ W

$141.8^2 \times 0.0375 = 754.02$ W

Total losses $= 1157.2$ W

Total power into the feeder $= 33.2$ A at a potential difference of 238 V $= 33.2 \times 238$

$= 7901.6$ W

and 141.8 A at a potential difference of 245 V $= 141.8 \times 245 = 34\,741$ W

Total power input $= 42\,642.6$ W

$$\text{Efficiency} = \frac{\text{output}}{\text{input}} = \frac{\text{input} - \text{losses}}{\text{input}}$$

$$= \frac{42\,642.6 - 1157.2}{42\,642.6}$$

$$= \mathbf{0.973}\ \textbf{p.u.}\ \mathbf{(97.3\%)}.$$

Problems on electricity supply

1   Calculate the magnitude of the current necessary to transmit a power of 100 kW single phase at a power factor of 0.7 lagging at (a) 500 V, (b) 5000 V. If in each case the conductor loop resistance is 0.2 Ω calculate the two power losses.

[(a) 285.78 A 16.33 kW, (b) 28.75A 163 W]

2   What are the main sources of energy used in the generation of electrical energy in the UK?

3   What problems with the burning of coal in power stations are overcome by the use of natural gas? What problems might become apparent to future generations because of this usage?

4   What is the function of a feed heater in a power station?

5   Why are generators in many major power stations connected directly to the 400 kV system? What savings may be made by developing local industrial generation with an input to the system at a lower voltage?

6   Compare the use of steel lattice towers and wooden poles for supporting overhead lines in electricity transmission and distribution systems.

7   What type of insulation is used in e.h.v. underground cables?

8   Define the terms (a) circuit breaker; (b) isolator; (c) switch.

9   Name three types of prime mover used to drive electricity generators.

10　Why is generation not carried out at the voltage level of major transmission?

11　When a circuit breaker interrupts a circuit an arc is drawn. How is this extinguished in an e.h.v. circuit breaker?

12　Why does the aluminium conductor used in overhead power transmission need a steel core?

13　What is meant by the term 'Safety Factor' as applied to overhead lines and supports?

14　What is the function of the stone-filled pit under an oil-filled transformer?

15　Why are large indoor substations often sectioned?

16　What is the function of a tap changer?

17　Why is a ring main used for power distribution wherever possible?

18　How are faults detected on overhead lines?

19　Why have accurate records to be kept of the route of underground cables?

20　Why is care taken to balance the loads across the phases of a three-phase system?

21　What is the name given to an electrically operated switch used to control motors?

22　List the advantages and disadvantages of an overhead transmission system as compared with one using underground cables.

23　The line voltage of a synchronous three-phase generator is 33 kV; what is the value of the voltage from one line to the star point?

[19 052 V]

24　A small, portable, three-phase generator operates at 415/240 V and supplies the following loads connected from the respective line to neutral:
Red phase, 10, 65 W fluorescent lamps operating at 0.6 power factor lagging,
Blue phase, electric kettle rated at 3 kW operating at unity power factor,
Yellow phase, woodworking equipment drawing 2 kW at 0.85 power factor lagging.
Calculate the magnitude and phase angle of the current returning to the generator in the neutral wire.

[$I_R = 4.51$ A at 0.6 lag, $I_Y = 12.5$ A at unity, $I_B = 9.8$ A at 0.85 lag. $I_N = 5.63$ A at an angle of $-125°$ from the red phase voltage.]

25　A three-phase motor has an electrical input of 500 kW at 11 kV (line) and operates at 0.85 power factor lagging. It is fed via a transformer from a 132 kV system. Assuming an ideal transformer, calculate (a) the line current supplied to the motor and (b) the line current at the input to the 132/11 kV transformer.

[(a) 30.87 A, (b) 2.57 A]

26   A three-phase, delta-connected induction motor draws a power of 23.5 kW from a 415/240 V supply. Its line current at this load is 38 A.
Calculate: (a) the operating power factor of the motor at this load, (b) the phase current of the motor.

[(a) 0.86 p.f., (b) 21.94 A]

27   A three-phase, delta/star transformer with line voltage ratio 11 000/415 V supplies a balanced load of 50 kW at 0.65 power factor lagging from its 415 V winding.
Calculate: (a) the secondary line current,
            (b) the primary line current,
            (c) the primary phase current; assuming ideal transformer operation.

[(a) 107 A, (b) 4.04 A, (c) 2.33 A]

28   A radial feeder *ABCD* is fed at 200 V. The loads are 20 A, 10 A and 10 A at *B*, *C* and *D* respectively. The loop resistances of the cable are:
AB = 0.1 Ω, BC = 0.15 Ω, CD = 0.05 Ω.
Calculate the efficiency of the system under these conditions.

[0.972 p.u.]

29   A feeder *ABCD* is fed at *A* and *D* at 220 V. A load of 20 A is situated at *B* which is 100 m from *A*. A load of 30 A is situated at *C* which is 120 m from *D*. The feeder is 420 m in length. The resistance of 100m of *single* conductor is 0.025 Ω. Determine the values of the currents in each section of the feeder and the minimum voltage.

[*AB* = 23.8 A; *BC* = 3.81 A; *CD* = 26.2 A; Min. p.d. at *C* = 218.43 V]

30   For the feeder shown in Figure 1.33 determine the power in the load which has the minimum potential difference across it.

[50 A load power = 11.794 kW]

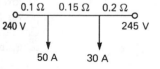

Figure 1.33

31   A ring main *ABCDEFGA* is fed at *A* at 250 V.
AB = 50 m, *BC* = 50 m, *CD* = 100 m, *DE* = 75 m, *EF* = 75 m, *FG* = 150 m and *GA* = 100 m.

A 100 m loop (go and return) of the ring has a resistance of 0.1 Ω. The loads are as follows:

$B = 20$ A, $C = 30$ A, $D = 10$ A, $E = 50$ A, $F = 20$ A, $G = 25$ A.

Determine the values of the currents in each section of the ring and the value of the minimum potential difference at a load.

[$AB = 89.6$ A; $BC = 69.6$ A; $CD = 39.6$ A; $DE = 29.6$ A; $EF = 20.4$ A; $FG = 40.4$ A; $GA = 65.4$ A; Min. p.d. at $E = 235.9$ V]

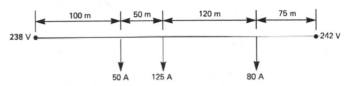

Figure 1.34

32 For the feeder shown in Figure 1.34 calculate (a) the individual load voltages and powers and (b) the efficiency of the system. A 100 m loop of the feeder has a resistance of 0.06 Ω.

[(a) 50 A load 231.75 V 11.587 kW, 125 A load 238.12 V 28.765 kW, 80 A load 235.22 V 18.817 kW; (b) efficiency = 59.169/61.293 = 0.963 p.u.]

Multiple choice questions

1 The conduction (copper) losses in a particular cable are 150 kW when the load transmitted is 200 MW. A load of 400 MW is transmitted through a second cable of the same material, each core being twice the cross-sectional area of those on the first cable. The conductor losses in the second cable will be:

  A 150 kW
  B 300 kW
  C 600 kW
  D 1200 kW.

2 Most extra-high voltage transformers for use in power supply systems have windings insulated using:

  A PVC
  B rubber
  C paper
  D PTFE.

3 Transmission voltages are made as high as practicable in order to:

  A minimize the size of the generators
  B reduce the cross-sectional area of the overhead transmission lines
  C allow circuit breakers to be more efficient since they are handling smaller currents
  D minimize the cost of underground cables in the system.

4   Overhead lines are often made of aluminium and employ a central steel core. The steel core is present:

   A   to increase the strength of the conductor
   B   to increase the overall conductivity of the conductor since the resistance of the steel is less than that of the aluminium
   C   to increase the circuit inductance
   D   to assist in the process of joining two sections of conductor together since it is difficult to weld aluminium.

5   A device for making a circuit dead which is not carrying current at the time but which may be used for energizing a circuit under any conditions is called:

   A   a circuit breaker
   B   a switch
   C   an isolator.

A device for making and breaking a circuit under any conditions is called:

   A   a contactor
   B   a circuit breaker
   C   an isolator.

6   The stone-filled pit under an oil-filled transformer is provided:

   A   since it is cheaper than a solid concrete floor
   B   it facilitates cable laying and renewal when required
   C   it assists in fire prevention and fire fighting.

7   A transformer tap changer allows:

   A   the oil to be changed periodically
   B   the output voltage to be maintained at the correct value
   C   the input voltage to be changed
   D   the input power factor to be changed.

8   A three-phase, four-wire system has a line voltage of 66 kV. Its phase voltage is:

   A   66 kV
   B   38.105 kV
   C   114.315 kV.

9   A two-wire system carries 50 A to a single load. The resistance of each wire is 0.05 Ω. The total conduction loss in the system is:

   A   2.5 W
   B   125 W
   C   250 W.

10 Each phase of a 440 V, three-phase, star-connected load carries 20 A at 0.7 power factor lagging. The current returning to the supply transformer star point is:

   A   87.5 A
   B   60 A
   C   zero
   D   28.6A.

11 A three-phase, four-wire system supplies a three-phase, star-connected load. The load is balanced and each phase draws 25 A. The resistance of each wire is 0.1 Ω. The total system losses are:

A   62.5 W
B   187.5 W
C   375 W.

12 The supply to electrical loads with a minimum of cable losses and the greatest security is achieved using:

A   a radial system
B   a feeder fed at both ends
C   a ring system.

2 Transformers

Summary

This chapter provides an introduction to the construction and performance of transformers. It looks at the operation of ideal transformers and investigates the laws and principles which determine their performance. It investigates the construction and applications of transformers together with their characteristics.

Faraday discovered that whenever a change in magnetic flux is associated with a coil of wire a voltage is induced in that coil. The value of the induced e.m.f. is proportional to the number of turns and the rate of change of magnetic flux in webers per second.

$$e = N\frac{\Delta\phi}{\Delta t} \text{ volts}$$

Alternating voltages of any desired value may be obtained by using the transformer which employs this principle. Voltages need to be changed between the points of generation and the consumer several times in order to arrive at the most economical levels for transmission and distribution. Generation is carried out at voltages between 11 kV and 25 kV whilst major transmission voltages are 275 kV and 400 kV in the UK. Domestic consumers are supplied at about 240 V.

Principle of action of the transformer

Figure 2.1 shows the general arrangement of a transformer with the secondary open-circuited. There are two coils, generally known as the primary and secondary, wound on an iron core. The iron core is made up of laminations which are about 0.3 mm thick. These have been rolled to the correct thickness, acid cleaned, polished and varnished or anodized on one side, and then made up into the correct core form.

When an alternating voltage V_p is applied to the primary a small magnetizing current flows which sets up a magnetic flux in the iron core. This alternating flux links with both the primary and secondary coils and with the iron inducing voltages in each of them. The voltage E_p induced in the primary coil opposes the primary applied voltage V_p according to Lenz's law. V_p is slightly greater than E_p and this difference causes a small current, known as the magnetizing current I_m, to flow in the primary. The voltage induced in the iron core causes eddy currents to flow so giving rise to the production of heat. There is also a heat loss caused by the alternating magnetization of the iron which is called the hysteresis loss. Both of these losses must be provided by the energy source to the primary winding and a current I_{H+E} flows in addition to I_m. These two currents sum to give I_o, the no-load current of the transformer. These losses will be dealt with further in a later section of this chapter. Finally the voltage E_s induced in the secondary winding is used to supply the load.

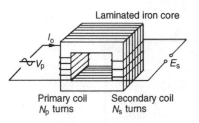

Primary coil N_p turns   Secondary coil N_s turns

Figure 2.1

EMF equation

Let the maximum value of the core flux be ϕ_m webers and the frequency f hertz.

The time taken for the flux to change from $+\phi_m$ to $-\phi_m$ is $\dfrac{\tau}{2}$ or $\dfrac{1}{2f}$ seconds. (Figure 2.2)

This is a change of $2\phi_m$ webers. Since $e = N\dfrac{\Delta\phi}{\Delta t}$ volts the average e.m.f. induced in the primary winding

$$= N_p \times 2\phi_m \div \frac{1}{2f} = 4N_p\phi_m f \text{ volts}$$

where N_p = number of turns on the primary winding.

For a sine wave, the r.m.s. value is 1.11 times the average value. Therefore $E_p = 4.44 N_p\phi_m f$ volts and if all the magnetic flux set up by the primary links with the secondary, $E_s = 4.44 N_s\phi_m f$ volts where N_s = number of turns on the secondary.

Therefore $\dfrac{E_p}{E_s} = \dfrac{N_p}{N_s}$

For a perfect transformer the input terminal voltage V_p may be regarded as being equal to the primary induced voltage E_p and the secondary output terminal voltage V_s equal to the induced voltage E_s.

Since total flux ϕ = flux density in webers per square metre B (tesla, T) × cross-sectional area A m^2 we may substitute $B_m A$ for ϕ_m in the e.m.f. equation.

$$E_p = 4.44 B_m A f N_p \text{ volts} \quad \text{and} \quad E_s = 4.44 B_m A f N_s \text{ volts}$$

Where A = cross-sectional area of the core in square metres.

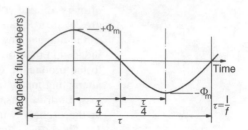

Figure 2.2

Current ratio

The product of current and turns is known as the magneto-motive force, m.m.f. Magneto-motive force sets up magnetic flux. The magnetizing current I_m flowing in the primary coil of N_p turns creates the core flux and e.m.f.s E_p and E_s as previously described. When a load is connected to the transformer secondary a current I_s flows creating a secondary m.m.f. of I_sN_s ampere turns. This creates a magnetic flux in the core which, according to Lenz's law, opposes that set up by the primary. E_p therefore falls, V_p is now much greater than E_p and there is a current inrush I'_s. The m.m.f. created by this current is I'_sN_p and this must balance I_sN_s.

I'_s is called the 'secondary current referred to the primary' or 'balancing current'. It is the additional current which flows in the primary because of the secondary loading.

We have therefore $I'_sN_p = I_sN_s$ or transposing:

$$\frac{I_s}{I'_s} = \frac{N_p}{N_s}$$

Notice that the current ratio is the inverse of the turns ratio, and hence of the voltage ratio. Increasing the voltage gives a reduction in current.

Test your knowledge 2.2

Complete the following table giving turns to the nearest integer and voltages and currents to the first decimal place.

	N_p turns	N_s turns	V_p volts	V_s volts	I_p amperes	I_s amperes
(a)	60	?	240	24	2	?
(b)	100	250	110	?	?	5
(c)	?	22	?	110	3.5	8

Example 2.1

A perfect transformer has 2000 primary turns and 500 secondary turns. The input voltage to the primary is 240 V. What is the turns ratio of the transformer? What is the value of the secondary voltage?

The turns ratio $= N_p/N_s = V_p/V_s = 2000/500 = 4$

Since there are more turns on the primary than on the secondary, the primary voltage is greater than the secondary voltage by a factor of 4.
Therefore the secondary voltage $= 240/4 = $ **60 V**

A transformer has full-load rating of 150 VA and a voltage ratio of 240 V:12 V.

(a) What are the values of primary and secondary currents on full load?
(b) What value of resistor connected to the secondary terminals would give the full rating in watts?

Example 2.2

A perfect transformer has 132 turns on its primary and has a voltage of 242 V applied to it. The output voltage is measured at 110 V.

(a) How many turns are there on the secondary?
(b) With the secondary connected to a load drawing 3.4 A, what is the value of the primary current?

Note that in a perfect transformer, or assuming ideal working, we ignore the core losses and magnetizing current when I_s' becomes the same thing as I_p.

(a) The turns ratio $N_p/N_s = $ the voltage ratio V_p/V_s
$= 242 \text{ V}/110 \text{ V} = 2.2$.
$N_p/N_s = 2.2$. Therefore $N_s = N_p/2.2 = 132/2.2$
$= 60$ turns.
(b) $I_s/I_p = N_p/N_s = 2.2$. Therefore $I_p = 3.4/2.2 = $ **1.55 A**

Example 2.3

A perfect transformer has its secondary connected to a load of 10 Ω. When the primary of the transformer is connected to a 240 V supply, 60 V is measured across the load. The primary has 80 turns. Calculate:

(a) the number of turns on the secondary
(b) the values of the primary and secondary currents.

(a) Turns ratio $= 240 \text{ V}/60 \text{ V} = 4$. The secondary voltage is less than the primary voltage so that the number of turns on the secondary is less than that on the primary.
Therefore, the number of secondary turns $= 80/4 = $ **20**.
(b) Using Ohm's law: $I_s = 60/10 = 6$ A.
Current ratio is inverse to the voltage ratio, therefore $I_p = 6/4 = $ **1.5 A**

The single-phase transformer

A single-phase transformer has two windings, a primary and a secondary, which are usually wound on a laminated iron core. The windings may be arranged as in Figure 2.1 when it is known as a core-type transformer. However, since it is important to cause as much as possible of the magnetic flux which is set up in the primary to pass through, or link with, the secondary, the windings are more commonly either arranged concentrically or interleaved as shown in Figure 2.3. The interleaved winding is sometimes referred to as a *pancake construction*. One type of core uses lamination stampings as shown in Figure 2.4.

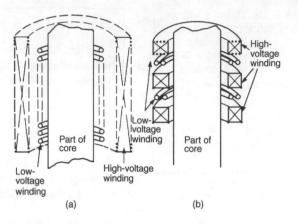

Figure 2.3

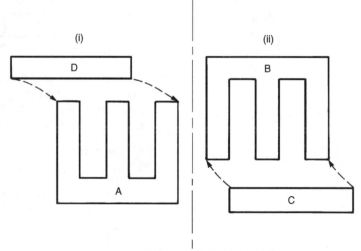

Add D to A, C to B and stack (ii) over (i)

Figure 2.4

The two 'E' stampings plus the two end-closing stampings form an element of the core. Many of these elements are added to make up the correct core thickness.

A complete air-cooled single-phase transformer is shown in Figure 2.5. This is known as a shell type. Even with the two coils so close together there is always some leakage flux, that is to say magnetic flux created by one of the coils which does not in fact link with the other. This occurs partly because, where the edges of the

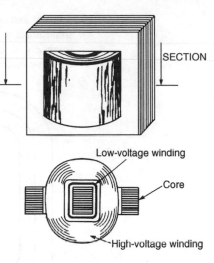

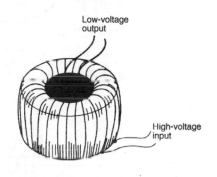

Figure 2.5

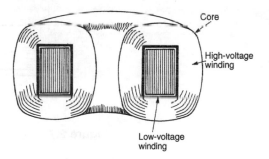

Figure 2.6

laminations meet, the grain structure of the iron is discontinuous; the shapes of the grains and the direction in which they have been rolled are different in one lamination from those in the lamination which it touches, unless immense care is taken when cutting the laminations from the large roll of strip material. This gives rise to an increase in reluctance at the joints, causing leakage flux and an increase in iron losses. Another reason for leakage flux is that, however the windings are arranged, there is a slight gap between primary and secondary and between windings and the core. Again, some leakage flux may be created. Leakage flux can create interference with other circuits, inducing voltages which cause hum in communication networks.

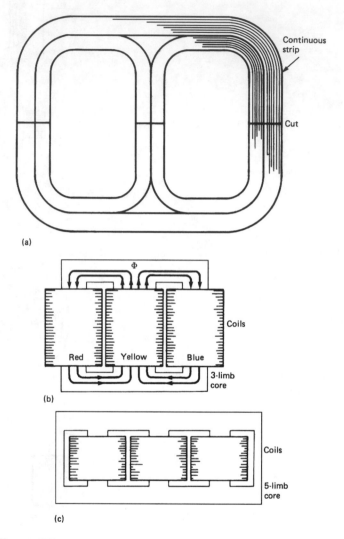

(a)

(b)

(c)

Figure 2.7

Where low core losses combined with virtually zero leakage flux are sought, a core formed from a continuous strip of grain-oriented steel may be used. The strip is wound on a mandrel of the correct size and subsequently heat treated to remove the stresses set up by this process. The windings are added either in the form of one of the arrangements already described or as shown in Figure 2.6 which shows the toroidal transformer as used in high-quality audio equipment. The path of the flux follows the line of the strip, there is no tendency to leak away and full use is made of the properties of the grain-oriented material. There are no joints across which the flux must pass. Another use of this construction is shown in the three-phase transformer in Figure 2.7.

Some typical uses for single-phase transformers are:

1   Power supplies for electronic equipment. Taking a supply at 240 V, the transformer will step this down to a lower value when it will be rectified and smoothed to give a direct supply to, for example, amplifiers, instruments and oscillators.

2   To provide a supply in workshops at reduced voltage for illumination and power tools. The reduced voltage is used for safety reasons and may be at 110 V or 25 V. The 25 V supply is often derived using a centre-tapped secondary. This point is earthed when the outputs are at ±12.5 V with respect to earth.

3   As an isolating transformer, when neither of the output wires is connected to earth. This means that a user touching either of the secondary wires would not receive an electric shock since there is nowhere for currents which would flow through the body to return to.

4   Electric arc welding. The primary is often connected across two lines of a three-phase supply and so is at 415 V. The secondary provides an output at between 90 V and 110 V when no welding is taking place. This falls to about 25 V once the arc is struck. This transformer is an instance of one in which very high leakage flux paths are present by design so that when load is taken from the secondary a collapse in voltage occurs. The use of two lines for the input spreads the load over two phases of the supply. Where several welders are in use they should not all be connected to the same two lines.

The three-phase transformer

Three-phase transformers are essentially three single-phase transformers wound on a single core. The windings are of concentric or pancake type. The windings are arranged on a laminated iron core with either three limbs or five limbs. They are shown in Figure 2.7. Notice that in Figure 2.7(a) continuous strip is used to form the yoke. When it is completely formed it is cut in half. This is called a 'cut core'. After carefully cleaning the cut and adding the windings, the two halves are reunited. Very little reluctance is added by the cutting and joining since at the join the grain all runs in the original direction, each lamination being virtually restored to its uncut state.

With the three-phase system, provided that the currents are balanced over the phases, the instantaneous sum of the currents

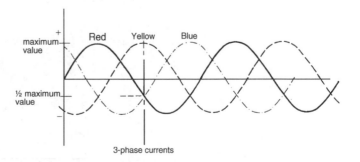

Figure 2.8

is zero. Now since each of the three currents produces a magnetic field, it follows that at any instant the sum of the magnetic fluxes will also be zero. In Figure 2.8 consider the instant at which the yellow current and hence the yellow flux are at their maximum positive values whilst each of the other two currents has one half of its maximum value, but negative. The fluxes in the red and blue limbs will have a magnitude one half that in the yellow limb and be in the opposite direction. In Figure 2.7(b) the maximum flux is shown upwards in the yellow limb; it splits into two halves, one shown downwards in the red limb and the other half downwards in the blue limb. At any instant this balance of fluxes will obtain.

Activity 2.1

Using Figure 2.8 select several other instants in time and demonstrate that at each of these instants the currents do indeed sum to zero. For the particular instant 30° after the yellow current has its maximum value, that is, as the blue current passes through its zero point, indicate on Figure 2.7(b) the directions and relative sizes of the fluxes.

If the phase currents are not equal, one of the limb fluxes will tend to be of greater magnitude than the sum of the other two. The out-of-balance flux will pass out of the three-limbed core and into the surrounding medium. This can be prevented by providing extra limbs. In the five-limb core shown in Figure 2.7(c) there are paths for out-of-balance flux. When the load is balanced the extra limbs have no effect; when there is out of balance these carry some flux so that none needs to leave the core.

Three-phase transformers are used at all stages of transmission and distribution to create the necessary voltages (see Chapter 1: The three-phase system and Transmission and distribution of electrical energy). Because of the necessity to reduce the transport weight of the very largest transformers, those associated with generators – stepping up from 25.6 kV to 400 kV – are often of the auto type. (See under 'The auto-transformer' below.)

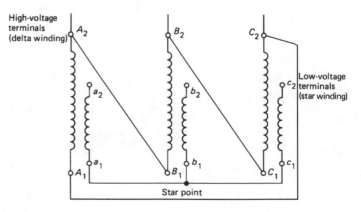

Figure 2.9

Three-limbed and five-limbed transformers are employed. All others are double wound, usually with three limbs.

The primary may be either star-connected or delta-connected but the secondary is almost invariably star-connected to provide an earthing point. Figure 2.9 shows a schematic arrangement of a delta-star three-phase transformer.

Transformer cooling

Transformers suffer eddy current and hysteresis losses in their cores and winding heating losses or copper losses in their windings as currents flow (I^2R). The heat generated must be removed or eventually the transformer will become hot enough to fail. For the smaller ratings, natural air cooling is sufficient. In such transformers the windings are often insulated using varnish and heat fairly readily passes through this into the surrounding air where it is convected away. Provided that free passage of air can be maintained, all will be well. Sometimes a fan is provided to give forced air cooling.

For higher ratings and in particular for higher voltages, the transformer windings are insulated using paper. The transformer is immersed in oil within a steel tank. The oil soaks the insulation and fills all the small gaps creating a uniform dielectric. In addition, the oil will carry away heat either by natural convection or by forced circulation by an external pump. The oil is cooled in one of three ways: (1) by natural air convection over external oil-filled tubes (Figure 2.10), (2) by forced air circulation using fans (Figure 2.11), or (3) as in (2) except that a water cooler is

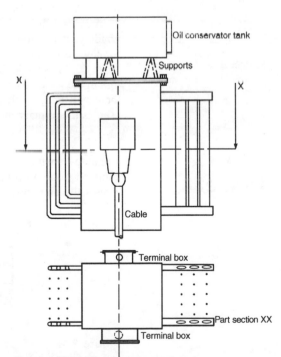

Two alternative cooling tube arrangements on an oil-natural air-cooled transformer

Figure 2.10

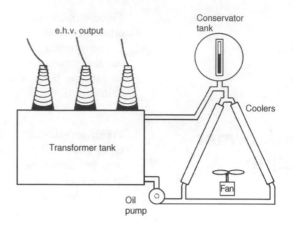

Figure 2.11

used, the water being drawn through many small tubes over which the oil is pumped. The water pressure is maintained lower than that of the oil so that any leakage is of oil into water which may be automatically detected, rather than water into oil which would cause the transformer to fail.

Transformer rating

A transformer has a name-plate rating which in effect informs us how much current the transformer windings can carry without overheating. The rating is quoted in volt-amperes (VA) at the full rated voltage. For example, a single-phase transformer with a rating of 10 kVA and a ratio of 500 V:100 V can carry $10\,000/500 =$ 20 A in its 500 V winding and $10\,000/100 = 100$ A in its 100 V winding. If the voltage is lower than specified on the name plate for any reason, the current may not be increased to give the original value of VA. Such an increase in current would increase the power loss in the windings and the transformer would overheat. Note that there will be a small additional current in whichever winding is used as the primary due to the magnetizing and core loss currents.

The auto-transformer

The auto-transformer has only one winding. Part of the winding is common to both primary and secondary which are therefore both electrically and mechanically linked.

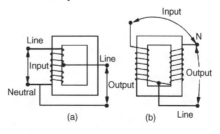

Figure 2.12

Figure 2.12 shows the possible arrangements of coils and core. The inputs and outputs are reversible providing for both voltage increase and decrease. Considering the transformer to be ideal, i.e. ignoring all losses, the simplified circuits shown in Figure 2.13 may be drawn.

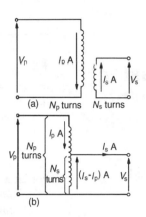

Figure 2.13

Figure 2.13(a) shows the current directions in the primary and secondary of a double-wound transformer. When the secondary is in fact part of the primary, the current in the secondary section becomes $(I_s - I_p)$ as shown in Figure 2.13(b) and the cross-sectional area of this section may be reduced so saving copper.

Example 2.4

A single-phase auto-transformer has a ratio of 500 V:400 V and supplies a load of 30 kVA at 400 V. Calculate the value of current in each section of the winding. Assume ideal operation.

$$30 \text{ kVA at } 500 \text{ V requires } \frac{30\,000}{500} = 60 \text{ A}$$

$$30 \text{ kVA at } 400 \text{ V requires } \frac{30\,000}{400} = 75 \text{ A}$$

Therefore, current in top section, $I_p = \textbf{60 A}$ and current in bottom (secondary) section, $(I_s - I_p) = (75 - 60) = \textbf{15 A}.$

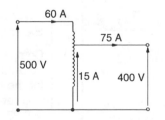

Figure 2.14

Observe that whereas in a double-wound transformer the secondary would be separate and have a cross-sectional area capable of carrying 75 A, with this construction it need only have a 15 A current-carrying capacity.

An expression for the saving in copper using the auto-transformer

The volume of copper in a winding is proportional to the number of turns and the cross-sectional area of the wire used, which in turn is proportional to the current to be carried. Therefore, volume of copper $\propto NI$.

For a double-wound transformer volume of copper $\propto N_p I_p + N_s I_s$ and assuming ideal operation $N_p I_p = N_s I_s$.

Therefore, volume of copper $= 2N_p I_p$.

For an auto-transformer (Figure 2.13)

$$\text{Volume of copper} \propto N_s(I_s - I_p) + (N_p - N_s)I_p$$

$$\propto N_s I_s + N_p I_p - I_p N_s - I_p N_s$$

$$\propto N_s I_s + N_p I_p - 2I_p N_p$$

and since $N_s I_s = N_p I_p$

$$\propto 2N_p I_p - 2I_p N_s.$$

$$\frac{\text{Volume of copper in the auto-transformer}}{\text{Volume of copper in a double-wound transformer}}$$

$$= \frac{2N_p I_p - 2I_p N_s}{2N_p I_p}$$

$$= 1 - \frac{N_s}{N_p}$$

Transposing:

$$\text{Volume of copper in the auto-transformer} = \left\{ 1 - \frac{N_s}{N_p} \right\}$$

$$\times \text{volume of copper in a double-wound transformer}$$

When V_s approaches V_p: a very small step up or step down: N_s approaches N_p when the saving in copper is greatest.

Test your knowledge 2.4

A single-phase auto-transformer has an input at 110 V and output at 440 V. It is rated at 5 kVA. Assuming ideal operation determine the value of currents in the two sections of the winding.

Example 2.5

Compare the volume of copper in a single-phase auto-transformer with that in a double-wound transformer for the ratio 400 V:300 V.

The ratio $= (1 - N_s/N_p) = (1 - 300/400) = (1 - 0.75) = 0.25$.

The volume of copper in the auto-transformer $= 0.25$ times that in the double-wound transformer.

Further advantages of the auto connection are:

1 The weight and volume of the auto-transformer are less.
2 The auto-transformer has a higher efficiency and suffers less voltage variation with changing load due to the better magnetic linkage between the primary and secondary.
3 A continuously variable output voltage is obtainable using the arrangement shown in Figure 2.15.

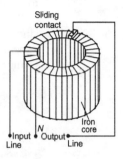

Figure 2.15

Auto-transformers are used to connect the 400 kV, 275 kV and 132 kV sections of the British grid system. When transporting very large high-voltage transformers by road the weight has to be carefully considered and quite often before major power system construction is commenced, roads and bridges have to be specially reinforced to carry such loads.

It is also worth noting that an increase in efficiency of only 0.5 per cent means a reduction in losses of 2500 kW in a 500 000 kW transformer.

Auto-transformers are also used to reduce the voltage supplied to induction motors during the starting period which reduces the potentially damaging current inrush during this period. Fluorescent lamps may be started using special auto-transformers designed to have high leakage flux. In this instance they are used to increase the voltage during the starting period. Once the lamp has struck the voltage collapses to a low value sufficient only to maintain the lamp discharge.

Disadvantages of the auto-transformer

1 Since the neutral connection is common to both primary and secondary, earthing the primary automatically earths the secondary. Double-wound transformers are sometimes used to isolate equipment from earth.
2 If the secondary suffers a short-circuit fault, the current which flows is very much larger than in the double-wound transformer due to the better magnetic flux linkage. There is more risk of damage to the transformer and circuit due to heating and the mechanical forces set up between current-carrying conductors.
3 A break in the secondary section of the winding stops the transformer action and the full primary voltage will be applied to the secondary circuit.

Current transformers

The current required to give full-scale deflection of a DC moving-coil ammeter is very small being typically only a few milliamperes. When large currents are to be measured a shunt or bypass resistor is used in conjunction with the meter. One method of using the shunt is shown in Figure 2.16. This method involves individual calibration of each such meter with its shunt to allow for the resistance of the connecting wires. Alternatively a moving-iron meter may be used. In either case the meter coil is at the potential of the circuit in which the current is being measured which gives rise to insurmountable problems with insulation at the higher voltages.

Where alternating currents are involved a shunt cannot be used since the proportion of the current which flows in the meter will depend on its impedance, which varies with frequency. A small change in frequency would upset the calibration of the meter.

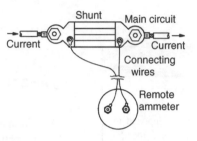

Figure 2.16

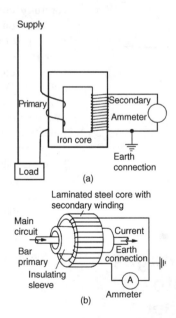

Figure 2.17

These problems are overcome by the use of current transformers which isolate the meter from the main circuit and allow the use of a standard range of meters giving full-scale deflections with 1, 2 or 5 A irrespective of the value of current in the main circuit. Two types of current transformer are shown in Figure 2.17.

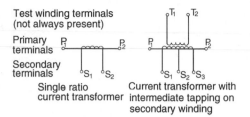

Test winding terminals (not always present)

Primary terminals

Secondary terminals

Single ratio current transformer

Current transformer with intermediate tapping on secondary winding

Figure 2.18

Figure 2.17(a) has a wound primary while Figure 2.17(b) shows the secondary wound on a core surrounding the main circuit conductor which is known as a bar primary. Figure 2.18 shows typical terminal markings for current transformers.

The primary of the current transformer is connected in series with the load on the circuit, replacing an ammeter, and has an extremely small voltage drop. The core flux is therefore small. (See e.m.f. equation.) The value of the primary current is determined entirely by the load in the main circuit and not by the load on its own secondary as in the case of the double-wound power transformer. The current transformer secondary load is an ammeter with a low impedance so that the terminal voltage is low, the effective load being typically between 2.5 VA and 30 VA.

Since the core flux is small, the core losses are small and the current transformer may be considered to be operating under ideal conditions when:

$$\frac{I_s}{I_p} = \frac{N_p}{N_s} \quad \text{Transposing: } I_s = I_p \frac{N_p}{N_s}$$

Example 2.6

A current transformer has a primary winding of 2 turns and a secondary winding of 100 turns. The secondary winding is connected to an ammeter with a resistance of 0.2 Ω. The resistance of the secondary of the current transformer is 0.3 Ω. The value of the current in the primary winding is 250 A.

Calculate: (a) the value of the current in the secondary,
(b) the potential difference across the ammeter terminals,
(c) the total load in VA on the CT secondary.

(a) $I_s = I_p \dfrac{N_p}{N_s} = 250 \times \dfrac{2}{100} = \textbf{5 A}$

(b) Potential difference across the ammeter terminals
$= I_s \times$ resistance of the meter
$= 5 \times 0.2$
$= \textbf{1 V}$

(c) Total resistance of the secondary circuit $= 0.2 + 0.3$
$= 0.5\ \Omega$

Since 5 A is flowing, the total induced e.m.f.
$$= 5 \times 0.5$$
$$= 2.5 \text{ V}$$
Total VA $= 2.5 \times 5$
$$= \mathbf{12.5 \ VA}$$

The secondary of a CT must never be open-circuited whilst the primary is carrying current since under these conditions $I_s N_s$ will be zero. In a power transformer this would cause I'_s to fall to zero. In the CT the primary current is determined only by the load in the main circuit and therefore does not fall when the CT secondary is open-circuited. The magnetic flux set up by the primary is now not balanced by that in the secondary and linking with the large number of turns on the secondary induces a very high voltage in it, large enough in some cases to destroy the insulation. Such high voltages can be a danger to life. In addition the unopposed primary core flux will give rise to hysteresis losses, not present when the flux and voltages were low, which will tend to overheat the core and windings causing further damage.

Voltage or potential transformers

A DC ammeter is converted into a voltmeter by the addition of a series resistor or multiplier which limits the current at full-rated voltage to that required to give full-scale deflection of the movement. Up to about 1000 V this arrangement is quite satisfactory. Insulating the meter movement, the terminals and the multiplier from the case and the panel in which the meter is situated presents no special problems. Above 1000 V hazards begin to present themselves. The cable to the meter may be long and vulnerable to damage. Insulation becomes difficult and increasingly expensive as voltages rise.

Where alternating voltages are to be measured, the voltage transformer is used to reduce the voltage at the meter to 63.5 V or 110 V typically. The voltage transformer is essentially a power transformer designed with a view to minimizing the core loss. Great care is taken to obtain maximum flux linkage between the coils and the winding resistance is made small by using conductors with a larger cross-sectional area than in a power transformer of similar rating. The secondary is used for measuring purposes only so that the current is small. The internal volt drops may generally be ignored so that

$$\frac{V_p}{V_s} = \frac{N_p}{N_s} \text{ (almost exactly)}$$

Figure 2.19 shows a voltage transformer connected to one phase of an 11 kV system. The transformer has a ratio of 100:1. The dial of the voltmeter is marked to indicate the voltage on the high-voltage side allowing for the transformer ratio.

Where voltage transformers are used to measure line voltages on a three-phase system the secondaries are star-connected giving line voltages of $63.5 \times \sqrt{3} = 110$ V at the meters.

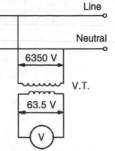

One phase of a 3-ph. system with
11 kV between lines

Figure 2.19

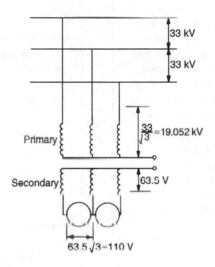

Figure 2.20

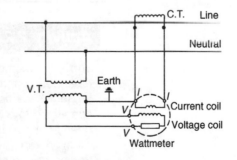

Figure 2.21

Measurement of power and power factor

Current and voltage transformers are used to isolate wattmeters from the high-voltage system in which the power is to be measured. The connections for a single-phase wattmeter are shown in Figure 2.21. The voltage and current coils are connected on one side to earth to prevent their potentials drifting up with respect to earth due to capacitance effects between their wiring and the high-voltage lines.

A load of 25 kVA with a power factor of 0.6 lagging is fed at 450 V from a single-phase supply. A voltage transformer with a ratio 4:1 and a current transformer with ratio 100:5, together with a wattmeter, are used to measure the power in the circuit. Determine:

(a) potential difference across the wattmeter voltage coil
(b) the current flowing in the wattmeter current coil
(c) the indicated power on the wattmeter.

Example 2.7

A voltage transformer of ratio 100:1 and a current transformer of ratio 100:5 are used to measure the power and power factor in a single-phase circuit using a wattmeter connected as shown in Figure 2.21. The potential difference across the wattmeter voltage coil is 63.5 V and the current in the current coil is 4.3 A. The wattmeter reading is 245 W.

Calculate for the primary circuit: (a) the current, (b) the phase voltage, (c) the power factor and (d) the power.

(a)　The CT ratio = 100:5.
　　　With 4.3 A in the secondary, the primary
　　　current $= 4.3 \times \dfrac{100}{5} = $ **86 A**.

(b)　The voltage transformer ratio = 100:1.
　　　Primary phase voltage $= 100 \times 63.5 = $ **6350 V**.

(c)　The power factor in the secondary is the same as that in the primary circuit assuming perfect transformers.

$$\text{Power factor} = \frac{\text{Power}}{\text{Volt-amperes}} = \frac{245}{63.5 \times 4.3} = \textbf{0.897}$$

(d)　Power in the primary $= VI \cos\phi$ watts
　　　　　　　　　　　　$= 6350 \times 86 \times 0.897$
　　　　　　　　　　　　$= \textbf{490 000 W}$

This is the same as secondary power $\times$ VT ratio $\times$ CT ratio

$$= 245 \times \frac{100}{1} \times \frac{100}{5} = 490\,000 \text{ W}$$

Activity 2.2

You are a member of a firm of consulting engineers. You have been approached by a farmer on a remote farm who presently has no electricity and who wants to be connected to a mains supply. In such situations this is often achieved by running one phase of an 11 kV supply on overhead poles to the farm, using a single-phase transformer to reduce the voltage to 240 V, feeding the farm through a circuit breaker to a distribution board and on into the farm buildings.

Write a report to the farmer explaining: (1) that he will need to have wooden poles erected on his land carrying the two bare conductors (one line and neutral); (2) there will be a pole-mounted transformer close to his buildings. Describe the transformer to him; its containment, what it will look like, how it will be cooled, what rating will be necessary (what equipment might he be using?).

(3) Supposing that he is required to pay for all the energy arriving at the farm, i.e. metering was required on the input side of the transformer to allow for the transformer losses, what additional equipment would be required? How would the scheme change if measurement on the 240 V side were permissible?

Activity 2.3

During your travels either locally or perhaps going on holiday you might well have seen a variety of overhead cables at a range of voltages from 240/415 up to 400 kV. Feeding them or being fed by them will be transformers of various constructions and ratings. Seek out any such transformer. Try to get some idea of its rating, its operating voltage, constructional features. Make a sketch of your transformer and write a report on your findings. ON NO ACCOUNT CLIMB INTO ANY RESTRICTED AREA TO DO THIS.

Further work on transformer losses

Core losses

(a) *Hysteresis*   Ferrous materials are made up of *domains* which are groups of atoms of the material. Each domain is a tiny magnet in its own right. In the non-magnetized state the domains form themselves into closed loops, north pole to south pole as one might imagine happening with a large number of permanent magnets thrown together in a box. Passing a current through a coil wound on a piece of ferrous material sets up a magneto-motive force which tends to break open the loops of domains causing them all generally to point in the same direction, along the axis of the coil. The core material exhibits strong magnetic effect at its two ends, a north pole at one end, a south pole at the other. Breaking open the loops and aligning the domains requires energy as would be the case in the permanent magnet analogy above. When the coil carries alternating current the domains have all to be oriented in one direction during a positive half-cycle and then in the opposite direction during the negative half-cycle.

The amount of energy involved depends on how many of the loops are broken and the extent that the domains are aligned. Externally this shows up as the strength of the magnetic flux density B tesla. The higher the external field strength, the more energy has been expended in re-orientating the domains. The energy used and the flux density are not proportional, however.

To sum up, the hysteresis loss is:

1 a function of the maximum working flux density B_m
2 directly proportional to the frequency. Since one reversal requires a certain amount of energy, performing the reversals at ever-increasing frequencies requires a directly proportional increase in energy input.

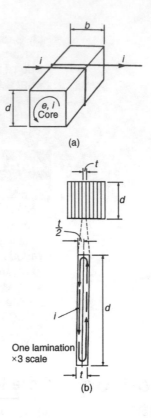

Figure 2.22

(b) *Eddy current loss*   Figure 2.22 shows a single turn wound on a solid conducting core. It is being fed from an alternating supply. An instant is chosen when the current in the coil is rising. The magnetic core flux is therefore rising and voltages are induced in the coil, in a secondary coil, if present, and since the core is made from conducting material, in the core itself. The e.m.f. induced in the core will drive a current which will generate heat. Core currents are known as eddy currents. The core is acting as a single turn and using $N = 1$ in the transformer e.m.f. equation the value of the induced e.m.f. is seen to be:

$$E = 4.44 \, B_{\mathrm{m}} A f \text{ volts}   \text{where } A = b \times d$$

The eddy current losses are reduced by rolling out the core material into very thin strips called laminations. Power transformer laminations are approximately 0.3 to 0.4 mm thick and are insulated from each other. Reducing the thickness of the material causes the large eddy current shown in Figure 2.22(a) to be broken up into many smaller paths as in Figure 2.22(b). Two effects are apparent: (1) the e.m.f. induced in each strip is less than in the solid core since the area per lamination is now $(t \times d)$ and (2) the resistance of the current paths is increased since they are much thinner. The ideal path for current is up one side of the lamination and down the other making the current path $t/2$ mm wide.

$$R = \frac{\rho \times \text{length of current path } (l)}{\text{area}}$$

where ρ = resistivity of the material.

The power loss in the core $= E^2/R$ watts where $E = $ e.m.f. induced in the lamination and $R = $ resistance of the current path.

The power loss in the lamination can therefore be expressed using an equation of the form:

$$\text{Power loss} = \frac{[4.44B_{\text{m}}(t \times d)f]^2}{\rho l/\text{area}} \text{ watts}$$

Since the actual current path cannot be predicted, we do not know the true area of the current path or that it will flow completely down one side (distance d) before returning up the other side. The known quantities are B_m, t, f and ρ. The other quantities are lumped into an experimentally determined constant. The important relationships to note are:

$$\text{Power loss } \propto B_{\text{m}}^2$$
$$\propto t^2$$
$$\propto f^2$$
$$\propto l/\rho$$

This means that halving the thickness of the lamination reduces the loss by a factor $(1/2)^2 = $ one quarter. Doubling the frequency of operation will increase the losses by $(2)^2 = $ four times. Increasing the resistivity of the material will reduce the losses in direct proportion to the increase. Adding between 0.5 and 4 per cent of silicon to the steel increases the resistivity but at the expense of the magnetic properties.

Working at high frequencies as in radio work, for example, means that different core constructions have to be employed. Laminations of nickel iron at thicknesses less than 0.1 mm may be used at frequencies up to about 10 MHz. For frequencies higher than this, since the losses are proportional to the square of the frequency, these may be considerable even with the thinnest laminations so that ferrite or dust cores are used in communications equipment up to about 150 MHz. Dust cores comprise iron, nickel-iron or molybdenum particles individually covered with an insulating material and held together in a resinous binder. Ferrites which are mixtures of iron oxides together with oxides of manganese, nickel and zinc have very high resistivities and are free from eddy current losses but since they saturate at lower flux densities than iron, a larger cross-section is required than when using grain-oriented steel. In ultra-high frequency applications air may have to be used as the core material since air suffers neither hysteresis nor eddy current losses.

Winding losses

Both primary and secondary windings have resistance and there will be associated power losses. Calling the primary and secondary currents I_p and I_s respectively and the corresponding resistances R_p and R_s the transformer copper losses will be:

$$I_p^2 R_p + I_s^2 R_s \text{ watts}$$

Winding leakage reactance

In practice it is impossible to ensure that all the magnetic flux created by the primary winding links with the secondary. Some of it is set up as shown in Figure 2.23.

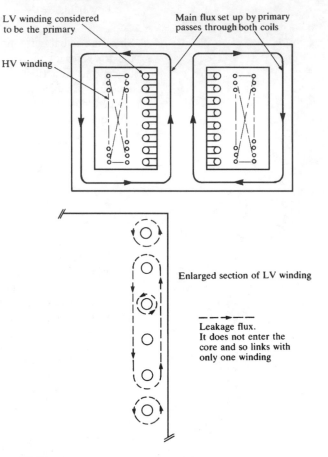

Figure 2.23

Similarly, when a current flows in the secondary this creates a flux which ideally ought to link with the primary because it is this linkage which lowers the impedance of the primary allowing extra current, known as balancing current, to flow in the primary. If there is leakage flux, the input current is smaller than ideal so limiting power output and the output voltage falls. In applications such as the welding transformer high leakage is built in so that the secondary voltage does fall as the arc is struck and the load current increases.

Transformer equivalent circuit

In order to be able to calculate the efficiency and output terminal voltage of a transformer on load we need an equivalent circuit diagram. At the heart of the diagram is a perfect transformer with no losses or imperfections. In a perfect transformer $V_p = E_p$ and $E_s = V_s$ and $\dfrac{E_p}{E_s} = \dfrac{N_p}{N_s}$

The resistances of the primary and secondary windings, R_p and R_s respectively, are shown external to the perfect transformer as shown in Figure 2.24.

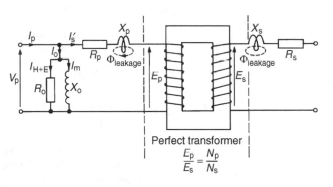

Figure 2.24

Leakage is represented as two small reactances X_p and X_s in the primary and secondary respectively. The combined value of these two components is determined experimentally in the short-circuit test. Considering Figure 2.23 it is seen that most of the flux created by the primary lies within the core and so links with both primary and secondary coils. However, some of the flux links only with the primary. Let us argue that 0.5 per cent of the total flux takes up leakage paths and that the primary winding has 200 turns. The flux entering the core is 99.5 per cent of that created by the primary winding. If the transformer were perfect and therefore the leakage flux did not exist, only 99.5 per cent of the 200 turns would be required, i.e. 199 turns. All the leakage flux is considered to be produced by the remaining single turn. In the equivalent circuit diagram the perfect transformer would be shown for this example as having 199 turns and the single turn producing only leakage (useless) flux drawn external to the transformer along with R_p. Since coils possess inductance and hence reactance, this external turn becomes X_p, the primary leakage reactance. The same argument can be applied to the secondary leading to the inclusion of X_s.

Open-circuit test

The normal rated voltage is applied to one of the windings of the transformer whilst the other is connected to a high-resistance voltmeter. The input voltage, power and current are measured using suitable-range instruments connected as shown in Figure 2.25.

Since the second winding is delivering virtually no current and the current in the winding being fed is extremely small, the total input power to the transformer as indicated on the wattmeter may be considered to be the hysteresis and eddy current losses of the core only. Using the results from this test we can construct the no-load phasor diagram and the no-load section of the equivalent circuit diagram for the transformer.

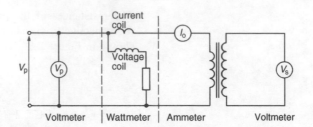

Voltmeter | Wattmeter | Ammeter   Voltmeter

Figure 2.25

Example 2.8

An open-circuit test on a 200 V:400 V, single-phase transformer gave the following results with the 200 V winding connected to a 200 V, 50 Hz supply.

Input current $= 2$ A   Input power $= 69.44$ W

Calculate: (a) the value of the iron loss current I_{H+E}
(b) the value of the magnetizing current I_m
(c) the no-load power factor and phase angle
(d) the values of shunt components which can be used in an equivalent circuit diagram to simulate these results.

Draw a phasor diagram to illustrate these results.

(a) The power input is regarded as core loss only. The core loss is therefore 69.44 W. It is being supplied from 200 V mains. Power can only be developed when a current flows which is in phase with the voltage.

$$\therefore 200 \times I_{H+E} = 69.44 \quad \text{so that}$$

$$I_{H+E} = 69.44/200 = \mathbf{0.374 \ A}$$

(b) The input current on no load is called I_o; in this example $I_o = 2$ A. The magnetizing current lags the applied voltage by exactly 90°.

$$\therefore \sqrt{(I_m^2 + I_{H+E}^2)} = \text{total input current}$$

$$= 2 \text{ A} \quad \text{(Pythagoras)}$$

$$I_m = \sqrt{(2^2 - 0.374^2)} = \mathbf{1.97 \ A}$$

(c) In a single-phase circuit $VI \cos\phi = $ power and $\cos\phi = $ power factor.

$$200 \times 2 \times \text{power factor} = 69.44$$

$$\text{Power factor} = \frac{69.44}{200 \times 2} = \mathbf{0.1736}$$

$$\phi = \text{INV} \cos 0.1736 = \mathbf{80°}$$

(a), (b) and (c) answers are shown in Figure 2.26.

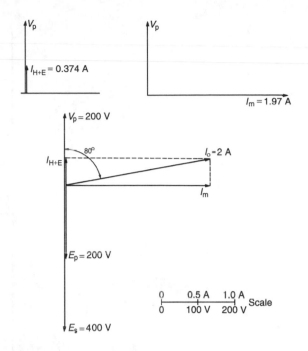

Figure 2.26

Test your knowledge 2.8

A 240 V:60 V, 50 Hz, single-phase transformer has a no-load current of 1.3 A at a power factor of 0.21 lagging, measured in the 240 V winding.

Calculate:
(a) the value of I_m
(b) the value of I_{H+E}
(c) the core iron losses
(d) the values of I_m and I_{H+E} in the 60 V winding if the transformer were turned round, being energized at 60 V, and the 240 V winding left open-circuited.

(d) The iron loss current being in phase with the applied voltage appears to be flowing in a pure resistance. This is called R_o.

$$R_o = \frac{200}{0.374} = 576 \ \Omega$$

The magnetizing current lags the applied voltage by 90° and hence appears to be flowing in a pure inductance. This is called X_o.

$$X_o = \frac{200}{1.97} = 101.52 \ \Omega$$

On no load the transformer draws the same current and power as a parallel combination of 576 Ω resistance and inductive reactance of 101.52 Ω. Since I_{H+E} and I_m are independent of load, R_o and X_o are shown in the equivalent circuit diagram as connected to the constant voltage input V_p. (See Figure 2.24.)

The loaded transformer

As we have already seen at the start of this section, the primary flux induces a back e.m.f. E_p which is very nearly equal to V_p and which opposes it according to Lenz's law. In Figure 2.26 we see $E_p = 200$ V drawn in phase opposition to V_p. Again, the primary flux also induces E_s so this is shown in phase with E_p.

The transformer is now loaded so that a current I_s flows and is lagging E_s by an angle ϕ_s.

The balancing current in the primary, $I'_s = I_s \dfrac{N_s}{N_p}$

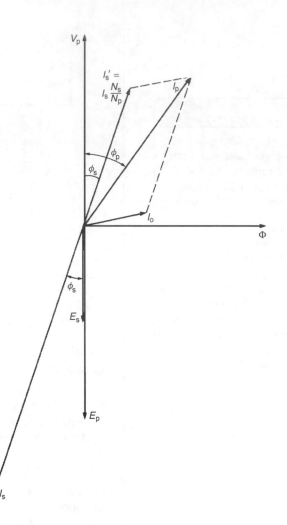

Figure 2.27

In Figure 2.27, I'_s is drawn to balance I_s and therefore lags V_p by the same angle as I_s lags E_s. Adding I'_s phasorially to I_o gives I_p, the total primary current. I_p lags V_p by an angle ϕ_p which is the operating power factor of the transformer for this particular load.

Activity 2.4

Perform the open-circuit test on a power transformer.

Connect the transformer to a supply of value equal to its name-plate rating as shown in Figure 2.25.

Compile a report describing the method and theory involved. Record the indicated values of applied voltage, current input and power. Calculate values for several quantities as investigated in Example 2.8 and develop the no-load phasor diagram for this transformer. Turn the transformer round and feed the other winding at its rated voltage whilst using the same instrumentation. Is the input current the same as in the first test? Is the input power the same as in the first test? Discuss and report on your findings.

Activity 2.5

Investigate the effect on transformer no-load losses of changing input voltage.

Connect a power transformer as shown in Figure 2.25 for the open-circuit test.

Measure the input power and I_o for input voltages: 20%, 40%, 60%, 80% and 100% of the name-plate value. Very gently increase the input voltage above the 100% value until I_o begins to increase rapidly. **Be very careful at this point because you are reaching core saturation**.

Evaluate values of I_m and I_{H+E} separately for 20%, 60% and 100% and the highest voltage you were able to reach.

Plot graphs of these against V_p, V_p on the vertical axis.

Does one of the two currents increase more rapidly than the other as the voltage increases? Why do you think this is so? Why is the relationship between I_m and I_{H+E} not linear?

Example 2.9

A single-phase transformer with a ratio of 500 V:100 V has a core loss of 200 W at full rated voltage. The magnetizing current is 2 A in the 500 V winding. The transformer delivers a current of 50 A from its 100 V winding at a power factor of 0.75 lagging. Determine:

(a) the iron loss component of current
(b) the no-load current of the transformer
(c) the balancing current in the primary winding
(d) the total primary current
(e) the power factor of the primary at this load.

(a) From Figure 2.28(a) it may be seen that

$$I_{H+E}/I_o = \cos \phi_o$$

$$\therefore I_{H+E} = I_o \cos \phi_o$$

But power input on no load $= V_p I_o \cos \phi_o$

$$\therefore 500 \times I_{H+E} = 200 \text{ so that } I_{H+E} = 200/500 = \textbf{0.4 A}$$

(b) $I_o = \sqrt{(I_{H+E}^2 + I_m^2)} = \sqrt{(0.4^2 + 2^2)} = 2.039 \text{ A}$

$$\cos \phi_o = \frac{I_{H+E}}{I_o} = \frac{0.4}{2.039} = \textbf{0.196} \quad \therefore \phi_o = \textbf{78.68}°$$

(c) $I'_s = 50 \times \dfrac{N_s}{N_p} = 50 \times \dfrac{E_s}{E_p} = 50 \times \dfrac{100}{500} = \textbf{10 A}$

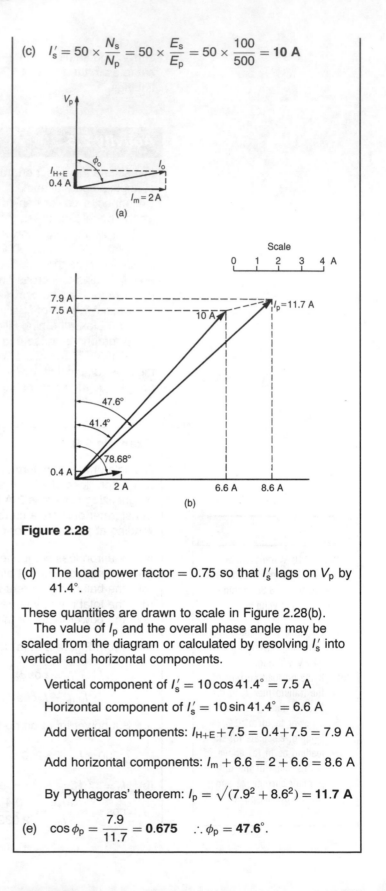

Figure 2.28

(d)   The load power factor $= 0.75$ so that I'_s lags on V_p by 41.4°.

These quantities are drawn to scale in Figure 2.28(b).
   The value of I_p and the overall phase angle may be scaled from the diagram or calculated by resolving I'_s into vertical and horizontal components.

   Vertical component of $I'_s = 10 \cos 41.4° = 7.5$ A

   Horizontal component of $I'_s = 10 \sin 41.4° = 6.6$ A

   Add vertical components: $I_{H+E} + 7.5 = 0.4 + 7.5 = 7.9$ A

   Add horizontal components: $I_m + 6.6 = 2 + 6.6 = 8.6$ A

   By Pythagoras' theorem: $I_p = \sqrt{(7.9^2 + 8.6^2)} = \textbf{11.7 A}$

(e) $\cos \phi_p = \dfrac{7.9}{11.7} = \textbf{0.675} \quad \therefore \phi_p = \textbf{47.6}°.$

The short-circuit test

One winding of the transformer is short-circuited through an ammeter while the other is fed from a variable voltage supply through a wattmeter and ammeter as shown in Figure 2.29. The input voltage is raised in increments until a value is reached at which currents of the order of the full-load value are circulating in both windings. Full load values are generally not exceeded from over-heating considerations. The voltage level to give these currents is usually in the region of 10–20 per cent of the normal rated voltage. Now since $E = 4.44B_m A f N$ volts, the flux will only be at 10–20 per cent of its working value. The hysteresis and eddy current losses are proportional to a power of B_m and so will both be extremely small. At 10 per cent of normal voltage the iron losses will be in the region of 1 per cent normal. The losses indicated on the wattmeter are therefore regarded as being due to the winding resistances only.

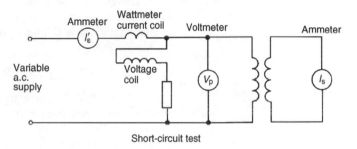

Figure 2.29

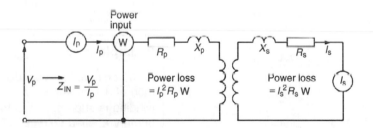

Figure 2.30

The series components for the equivalent circuit may be deduced from the short-circuit test results in the following manner with reference to Figure 2.30.

Power input as indicated on the wattmeter, $P = I_p^2 R_p + I_s^2 R_s$ watts

$$I_p N_p = I_s N_s \quad \therefore I_s = I_p N_p / N_s$$

Hence $P = I_p^2 R_p + [I_p N_p / N_s]^2 R_s$

$$= I_p^2 \{R_p + [N_p / N_s]^2 R_s\} = I_p^2 R_t^p$$

The effective resistance as seen by the input voltage on the primary side is given by:

$$R_t^{primary} = \{R_p + [N_p / N_s]^2 R_s\} \ \Omega$$

From the short-circuit test:

$$\frac{\text{Power input}}{I_p^2} = \{R_p + [N_p/N_s]^2 R_s\} = R_T^P$$

Where R = resistance, T = total, P or primary = as seen on the primary side.

Generally we only know total values; we never need to know separate values for the primary and secondary.

$$\frac{V_p}{I_p} = \text{input impedance}, Z_T^{\text{primary}} \text{ or } Z_T^P$$

and since $\sqrt{(R^2 + X^2)} = Z$ generally, $\sqrt{(Z^2 - R^2)} = X$
and in particular: $\sqrt{[Z_T^P]^2 - [R_T^P]^2} = X_T^P$
X_T^P = total leakage reactance of the transformer as seen by the primary.

Bringing the secondary quantities into the primary using the turns ratio squared is known as referring the secondary values into the primary. It is similarly possible to refer the primary values into the secondary.

$$R_T^S = \{R_s + [N_s/N_p]^2 R_p\} \qquad X_T^S = \{X_s + [N_s/N_p]^2 X_p\}$$

The method is demonstrated in Example 2.10. Note that when referring values from a low voltage winding into a higher voltage winding the values of resistance and reactance get larger since in the higher-voltage winding the currents are smaller and to give the same effect as in the lower-voltage winding the components need to be larger. Conversely when moving into a lower-voltage winding the currents are larger and the components must be smaller.

Example 2.10

A short-circuit test was done on a transformer with a turns ratio of 3:1 as shown in Figure 2.30. The low-voltage winding is short-circuited through an ammeter. The primary applied voltage = 50 V. The primary input current = 100 A. The power input = 1987 W.

(a)   Determine the values of (i) Z_T^P, (ii) R_T^P, (iii) X_T^P.
(b)   Using the turns ratio, refer the values in (i), (ii) and (iii) above into the secondary so finding (iv) Z_T^S, (v) R_T^S and (vi) X_T^S.
(c)   Verify that the calculated value of R_T^S gives the same value of winding power loss as quoted in the question.

(a)   (i) $Z_T^P = V_p/I_p = 50\,\text{V}/100\,\text{A} = \mathbf{0.5\ \Omega}$
       (ii) $R_T^P = \text{power loss}/I_p^2 = 1987/100^2 = \mathbf{0.1987\ \Omega}$
       (iii) $X_T^P = \sqrt{(0.5^2 - 0.1987^2)} = \mathbf{0.4588\ \Omega}$.

(b)   The turns ratio is 3:1. Referring into the lower-voltage winding means using $[N_s/N_p]^2 = [1/3]^2 = 1/9$ (the values will be smaller).

       (iv) $Z_T^S = 0.5 \times 1/9 = \mathbf{0.0555\ \Omega}$

(v) $R_T{}^S = 0.1987 \times 1/9 = \mathbf{0.0221}\ \Omega$

(vi) $X_T{}^S = 0.4588 \times 1/9 = \mathbf{0.051}\ \Omega$.

(c) $I_s = I_p N_p / N_s = 100 \times 3/1 = 300\ A$

Power losses using secondary quantities I_s and $R_T{}^S$

$$P = 300^2 \times 0.0221 = \mathbf{1987\ W}\ \text{(closely, since the}$$
$$\text{resistance value was slightly truncated).}$$

Activity 2.6

Perform a short-circuit test on a power transformer as described in this section. From it determine the values of total impedance, reactance and resistance referred both to the primary and secondary windings. It would be advantageous if the transformer used were the same one used in the open-circuit test in Activity 2.4. If it is not, then measure the core losses of this transformer as described for the open-circuit test and record all the data with a view to calculating its efficiency after the next section.

Transformer efficiency

The losses in a transformer are the eddy current and hysteresis losses which are measured using the open-circuit test, and the conduction or copper losses which are determined in the short circuit test.

$$\text{Efficiency} = \frac{\text{Output power}}{\text{Input power}} = \frac{\text{Output power}}{\text{Output power} + \text{losses}}$$

$$= \frac{\text{Output VA} \times \text{power factor}}{\text{Output VA} \times \text{power factor} + \text{core loss} + \text{winding loss}}$$

Example 2.11

A 240 V:110 V transformer is rated at 15 kVA. On an open-circuit test at full-rated voltage the power input was 400 W. On a short-circuit test the power input was 540 W at full-rated current.

Determine: (a) $R_T{}^P$ regarding the 240 V winding as the primary.

           (b)   The efficiency of the transformer for each of the following loads: (i) full load at unity power factor, (ii) full load at 0.6 power factor lagging and (iii) one half-full load at unity power factor.

(a)   Full rated current in the 240 V winding $= \dfrac{15\,000}{240}$

      $= 62.5\ A$.

$$I_p^2 R_T^P = \text{total copper loss in the transformer} = 540 \text{ W}$$

$$62.5^2 \times R_T^P = 540 \quad R_T^P = \frac{540}{62.5^2}$$

$$= 0.1382 \ \Omega$$

(b) (i) Efficiency =

$$\frac{15\,000 \times 1 \ (\text{unity power factor})}{15\,000 + 400 \ (\text{core loss}) + 540 \ (\text{copper loss on full load})}$$

$$= \mathbf{0.94}$$

(ii) Efficiency = $\dfrac{15\,000 \times 0.6}{15\,000 \times 0.6 + 400 + 540} = \mathbf{0.905}$

(iii) On half load, the current in the 240 V winding = 62.5/2 = 31.25 A.

Copper losses = $31.25^2 \times 0.1382 = 135$ W

(Half the current, one quarter the copper losses)

Efficiency = $\dfrac{7500 \times 1}{7500 + 400 + 135} = \mathbf{0.93}$

Example 2.12

A 480 V:240 V, 20 kVA single-phase transformer has a core loss of 200 W and a copper loss of 45 W when one half of full-rated current flows in the windings. Calculate the efficiency of the transformer when delivering:

(a) 10 kVA at 0.8 power factor lagging
(b) 20 kVA at unity power factor
(c) 15 kVA at 0.75 power factor lagging.

Half-rated current in the 480 V winding = 10 000/480
$$= 20.83 \text{ A.}$$

$R_T^P = 45/20.83^2 = 0.1037 \ \Omega$ (480 V winding regarded as primary)

(a) Efficiency = $\dfrac{10\,000 \times 0.8}{10\,000 \times 0.8 + 200 + 45} = \mathbf{0.97}$

(b) Efficiency = $\dfrac{20\,000}{20\,000 + 200 + 41.66^2 \times 0.1037} = \mathbf{0.98}$

(c) Efficiency = $\dfrac{15\,000 \times 0.75}{15\,000 \times 0.75 + 200 + 31.24^2 \times 0.1037}$

$$= \mathbf{0.974}$$

Test your knowledge 2.10

A 440 V:110 V transformer has a rating of 50 kVA. It has a total winding resistance referred to the primary, R_T^P, of 0.15 Ω. (Primary = 440 V winding.) The core loss at full working voltage = 2 kW. The secondary supplies its full rating at 0.85 power factor lagging. What is the efficiency of the transformer under these conditions?

Transformer regulation

When a transformer is loaded its output voltage changes. The change in voltage expressed as a percentage of its no load value is defined as the regulation of the transformer.

On no load, since there is no output current, there are no voltages developed across R_p, X_p, R_s or X_s. This means that $V_p = E_p$ and $V_s = E_s$. Loading the transformer causes currents to flow in all the series elements mentioned and the output voltage will change.

Percentage regulation

$$= \frac{\text{Change in output terminal voltage}}{\text{No-load output voltage } E_s} \times 100\%$$

$$= \frac{E_s \text{ (no load)} - V_s \text{ (measured at a specified load)}}{E_s \text{ (no load)}} \times 100\%$$

Example 2.13

A transformer has a no-load output voltage of 240 V. When carrying full load at 0.8 power factor lagging the transformer regulation is 4 per cent. What is the output terminal voltage at full load at this power factor?

$$\text{Percentage regulation} = \frac{\text{Change in terminal voltage}}{\text{No-load output voltage}} \times 100$$

$$4 = \frac{\text{Change in terminal voltage}}{240} \times 100$$

Transpose: Change in terminal voltage $= \dfrac{4 \times 240}{100} = 9.6$ V

$\therefore$ Output terminal voltage $= 240 - 9.6 = \mathbf{230.4 \ V}$

Activity 2.8

Investigate the effect of power factor on the regulation of a transformer. Set up a double-wound transformer as in Activity 2.4 with an ammeter and voltmeter in the primary and a sensitive voltmeter across the secondary output terminals. Using a resistive load, set as near as possible full-load current in the primary whilst observing the effect on the secondary voltage. If the effect of loading is to pull down the primary input voltage, take this into account.

Calculate the regulation of the transformer for this load.

What is the effect of varying the power factor? Using an inductive loading bank try to obtain the same current levels (closely) for a range of lagging power factors. What happens to the regulations as the power factor decreases?

Try now connecting any capacitive load to the secondary. It may not be possible to match the previous currents but any degree of loading should show up the effect. What happens to the output voltage when a capacitive load is connected? In areas with long distances of underground cables which are highly capacitive, at night when the connected load falls to a low value, this capacitance can have a pronounced effect with a change in voltage taking place as you have observed.

Condition for maximum efficiency

The maximum efficiency of a transformer occurs at a load at which the winding copper loss is equal to the core loss.

Example 2.14

A single-phase transformer has a core loss of 150 W and a total referred resistance in the secondary of 0.23 Ω. Calculate:

(a) The value of current in the secondary which will cause the transformer to operate at its maximum efficiency.
(b) The value of this maximum efficiency if the output terminal voltage when supplying the current in (a) above at 0.85 power factor lagging is 237 V.
(c) The efficiency of the transformer at the same power factor when supplying one quarter of the value of current in (a) above assuming that the terminal voltage does not change significantly from its value in (b).

(a) $I_s^2 R_T^S$ = total copper loss in the transformer. For maximum efficiency this must equal the core loss.

$\therefore I_s^2 \times 0.23 = 150 \quad I_s = \sqrt{(150/0.23)} = \mathbf{25.54 \ A}$

(b) Efficiency

$$= \frac{\text{Output}}{\text{Output} + \text{core loss} + \text{winding loss}} \times 100\%$$

$$= \frac{237 \times 25.54 \times 0.85}{237 \times 25.54 \times 0.85 + 150 + 150} \times 100\%$$

(Core and winding loss equal for maximum efficiency)

$= \mathbf{94.5\%}$

(c) One quarter of the current = 6.385 A.

Copper loss = $6.385^2 \times 0.23 = 9.38$ W.

Efficiency

$$= \frac{237 \times 6.385 \times 0.85}{237 \times 6.385 \times 0.85 + 150 + 9.38} \times 100\%$$

$= \mathbf{88.9\%}$

Test your knowledge 2.11

A transformer with voltage ratio 440:240 has a core loss at full working voltage of 96 watts. The total resistance of the transformer R_T^P in the 440 volt winding is 0.13 Ω. Determine:

(a) the value of current in the 440 V winding to give maximum efficiency
(b) the value of this efficiency.

Maximum power transfer

When considering equipment supplied with power from the mains supply it is never necessary to consider the conditions for maximum power transfer. The power supply is treated as being infinitely great compared with the demand being made, and a load to suit the user rather than to suit the supply network is connected.

However, when a small source of power is involved, such as the output from an audio amplifier or a telephone system, it is often important that the load be chosen of the correct magnitude in order that the maximum use be made of the available power from the source. The ideal load will be resistive since with inductive components, although there may be adequate voltage and current, the power will be determined by the power factor of the circuit.

Maximum possible power transfer from a source with internal impedance Z_I into a load occurs when the load has a resistance equal in magnitude to Z_I.

Matching may be achieved by using passive networks or by the use of transformers.

For the transformer method, consider the case with a load with pure resistance R_L (L = load). We have seen that a resistance R_S in the secondary of a transformer appears as a referred value $[N_p/N_s]^2 R_s$ in the primary. A load of R_L Ω connected to the secondary terminals of the transformer may be referred into the primary in the same manner.

Given that the internal resistance of the voltage source connected to the primary – the amplifier, telephone system etc. – is Z_I then for maximum transfer of power:

$$Z_I = [N_p/N_s]^2 \times R_L$$

Example 2.15

The output stage of an amplifier has an internal resistance of 512 Ω and this is to supply a load with resistance 8 Ω. The amplifier develops an internal e.m.f. of E volts. Determine

(a) the value of load power with the amplifier connected directly to the load,
(b) the turns ratio of an ideal transformer to give perfect matching between source and load,
(c) the power developed in the load when so matched, and
(d) the ratio of the matched/unmatched powers.

(a) Power $= I^2 R_L$ and $I = E/(R_I + R_L)$
$$= E/(512 + 8) = E/520$$
Power $= [E/520]^2 \times 8 = \mathbf{29.6}E \ \mu\mathbf{W}$

(b) $512 = [N_p/N_s]^2 \times 8$ $[N_p/N_s]^2 = 512/8 = 64$
$N_p/N_s = \sqrt{64} = \mathbf{8}$

(c) Now that the load resistance appears to be equal to the source resistance the total effective resistance $= 512 + 512 = 1024 \ \Omega$.

$$I = E/1024$$

$$\text{Load power} = [E/1024]^2 \times 512 = \mathbf{488.3}E \ \mu\mathbf{W}$$

(d) Ratio of the powers $= 488.3/29.6 = \mathbf{16.5}$

By matching, the load power has been increased by 16.5 times over the unmatched state.

Activity 2.9

Now that you have worked through the chapter and have performed various tests becoming familiar with transformer characteristics, select one type of transformer – power, current, voltage or other – and research the range available. What ratios are there, what cooling arrangements do they have, how large are they, what are their terminal arrangements etc.? Information is available in libraries, from suppliers and possibly manufacturers. Prepare and give a presentation to your group using written and visual material. There are film and video libraries from whom material might be borrowed to help such a presentation.

Problems on transformers

1   Describe the action of the transformer whereby a voltage appears at the secondary terminals of an energized transformer.

2 What is m.m.f.?

3 What are the advantages of forming a transformer core from a continuous strip of cold-rolled grain-oriented steel as compared with building it up from lamination stampings of the same material?

4 Define the term 'transformer rating'.

5 What alternative materials to laminated steel are there for transformer core operating at very high frequencies?

6 What is meant by the term 'leakage flux'?

7 What are the factors which influence eddy current losses in a transformer core?

8 What is hysteresis loss? What is the effect of loading a transformer on this loss?

9 What is 'copper loss' in a transformer? How does it vary as the loading is increased?

10 A transformer has a copper loss of 50 W when carrying one half-full load. Calculate the value of copper loss at (a) 70 per cent full load (b) full load.

[(a) 98 W, (b) 200 W]

11 A transformer has an iron loss of 100 W. What will be the value of the copper loss at maximum efficiency?

[100 W]

12 Calculate the efficiency of a single-phase transformer on full load at 0.8 power factor lagging given the following data.
Short-circuit test results. 100 V winding short-circuited.
Power input to 500 V winding = 35 W. Current in 500 V winding = 20 A.
Open-circuit test results. 500 V winding open-circuited.
100 V applied to the 100 V winding.
Input = 5 A at 0.2 power factor lagging.

[0.985]

13 A short-circuit test on a 5 kVA, 1730:440 V single-phase transformer yielded the following results:
With full-load current in the secondary (440 V) winding, power input to the primary = 106 W at 125 V.
(a) Determine the values of (i) R_T^P, (ii) R_T^S.
(b) If the core loss of this transformer is 86 W, determine the operating efficiency for one half-full load at 0.8 power factor lagging.
(c) (i) At what load should this transformer operate to achieve its maximum efficiency?
(ii) The value of this efficiency assuming operation at unity power factor.

[(a) (i) 12.69 Ω (ii) 0.82 Ω, (b) 0.947 p.u., (c) (i) 4504 VA (ii) 0.96 p.u.]

14   A 50 kVA, 11 kV:240 V single-phase transformer draws 1.2 A at 0.3 power factor lagging from the 11 kV supply with the 240 V winding open-circuited. For the transformer on full load at 0.75 power factor lagging, calculate: (a) the secondary current, (b) the primary balancing current, (c) the primary power factor and (d) the primary current.

[(a) 208.33 A, (b) 4.545 A, (c) 0.673, (d) 5.6 A]

15   Describe three ways in which transformers are cooled.

16   Why is the secondary winding of a three-phase transformer almost invariably star-connected?

17   What is the function of a conservator tank as fitted to a power transformer?

18   Why are some very large transformers auto-connected?

19   A transformer has a secondary output voltage on no load of 240 V. On full load this falls to 236 V. What is the value of the transformer regulation?

[1.66%]

20   Give two advantages and two disadvantages of the auto connection.

21   Estimate the ratio of copper in an auto-transformer to that in a double-wound transformer for ratios (a) 400 kV:132 kV and (b) 400 kV:275 kV.

[(a) 0.67, (b) 0.313]

22   A 10 kVA transformer has a full-load copper loss of 850 W and a core loss of 650 W. Determine: (a) what proportion of full load on the transformer will result in it operating with maximum efficiency and (b) what is the efficiency at this load?

[(a) 0.8744 times full load, (b) 0.858 p.u.]

23   What is an isolating transformer? Where would they be used?

24   Explain the possible effects of open-circuiting the secondary of a current transformer whilst current is flowing in the primary circuit.

25   What is the function of a voltage transformer? How does its construction vary from that of a normal power transformer?

26   Why is one terminal of a CT secondary connected to earth?

27   Explain why it is usual to use both current and voltage transformers when measuring the power in an h.v. circuit.

28   What is the function of a matching transformer? How would the performance of a loudspeaker be affected if incorrect matching were employed?

29 A load of 20 kVA is fed at 240 V from a single-phase supply. A current transformer with ratio 80:5 and a voltage transformer with ratio 5:1 are used to drive a wattmeter which indicates 200 W. Determine: (a) the current in the wattmeter current coil, (b) the voltage on the wattmeter voltage coil and (c) the load power factor.

[(a) 5.208 A, (b) 48 V, (c) 0.8 power factor]

30 A current transformer has a bar primary and 200 turns on its secondary. The secondary is connected to an ammeter with resistance 0.25 Ω. The resistance of the CT secondary winding is 0.35 Ω. The current in the bar primary is 850 A. Determine: (a) the current in the secondary, (b) the potential difference across the ammeter and (c) total load in VA on the secondary.

[(a) 4.25 A, (b) 1.063 V, (c) 10.84 VA]

Multiple choice questions

1 A transformer core is laminated:

A to reduce the hysteresis loss
B to reduce the eddy current loss
C to make the core lighter than if made from solid material
D to facilitate the cooling of the core.

2 Cold-rolled grain-oriented steel is used in transformer cores because:

A it gives a lower magnetic flux and hence lower leakage than other steels
B it is cheaper than any of the other materials available
C it can be magnetized much more easily than mild steel for a given input magnetizing current which has the effect of reducing the losses
D it can be assembled more or less at random with sections from sheets of the material, not having to take care to align the grain.

3 A single-phase transformer supplies a current of 10 A to a load from its secondary. The primary is fed at 240 V and the input current is 2.5 A. The secondary voltage is:

A 60 V
B 960 V
C 120 V.

4 A double-wound transformer of ratio 440:240 V contains a mass of 50 kg of copper in its two windings. An auto-transformer of the same rating would contain (closely):

A 50 kg
B 100 kg
C 18 kg
D 25 kg of copper.

5 Auto-transformers are used to interconnect the 400 kV and 132 kV transmission systems in the UK since:

    A double-wound transformers are unsuitable
    B auto-transformers are more reliable than double-wound transformers
    C auto-transformers weigh less than double-wound transformers of similar rating and ratio
    D auto-transformers deliver less current into a short-circuit fault should one occur.

6 Should the secondary section of the winding in a step-down auto-transformer fracture, the voltage at the secondary terminals would:

    A rise
    B fall
    C remain substantially unchanged.

7 A current transformer is used when measuring:

    A alternating currents
    B direct currents
        it has advantages over the shunt in that:
    C it is cheaper
    D it isolates the measuring instrument from the main circuit
    E the accuracy is not affected by small changes in frequency
    F the meter may be connected to and disconnected from the current transformer with no special precautions since the output voltage from the current transformer is very low.
        Select A or B together with *two* from C to F to make a correct statement.

8 Single-phase power is measured using a current transformer with a ratio 50:5 A and a voltage transformer with ratio 3300:110 V. The wattmeter indicates 300 W. The power in the primary circuit is:

    A 3 kW
    B 90 kW
    C 9 kW
    D 12 kW.

3

DC machines

Summary

This chapter describes the construction of a DC machine and explains the difference between the lap and wave windings. The action of the commutator is examined and the e.m.f. equation is derived. The interaction between pole magnetic flux and armature current which produces torque is examined. The necessity for and the action of a DC motor starter are discussed. The difference between armature reaction and reactance voltage is explained. The reversal of armature current direction as associated with the operation of the machine as a generator, losses and the power flow diagram of DC machines are examined.

Principle of action

From Faraday's original work we know that $e = \Delta\phi/\Delta t$ volts for a single turn where e = electromotive force in volts, $\Delta\phi$ = flux linked in webers and Δt = time interval in seconds.

Figure 3.1 shows a conductor of length l metres moving through an area of uniform magnetic field. The total flux is ϕ webers and the conductor moves a distance of d metres at a uniform velocity of v metres per second.

$$\text{Time taken} = \frac{\text{distance}}{\text{velocity}} = \frac{d}{v} \text{ seconds}$$

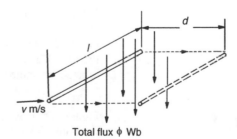

Total flux ϕ Wb

Figure 3.1

$$\text{Total flux linked} = \text{flux density} \times \text{area of the field}$$
$$= B \times l \times d$$

$$e = \frac{\text{flux linked}}{\text{time}} = \frac{B \times l \times d}{d/v} = Blv \text{ volts}$$

If the ends of the conductor are joined to an external circuit i amperes will flow. Power $= ei$ watts. Work done $=$ power $\times$ time $= ei \times$ time.

Therefore, work done $= \left\{ \dfrac{\text{Flux linked}}{\text{time}} \right\} \times i \times \text{time} = Bldi$ joules

But work done $=$ force $\times$ distance operated through $=$ force $F \times d$

$$Bldi = Fd \qquad \text{and} \qquad F = Bli \text{ newtons}$$

Thus, to generate an e.m.f. the requirement is for magnetic flux, a conductor and relative motion between them. To produce a force on a conductor the requirement is for magnetic flux and a current flowing in the conductor.

In generators the flux is usually provided by a system of electromagnets and the motion of the conductors provided using an engine.

In motors the current is provided from an external source. This current both excites the electromagnets and feeds the conductor system.

In the DC motor the force created by the current flowing in the conductor system causes them to accelerate if they are free to move. The motion of the conductors through the field causes an e.m.f. to be generated. By Lenz's law, the generated e.m.f. opposes the supply voltage. For this reason it is called the back e.m.f. At a certain speed dependent upon the mechanical forces involved, the strength of the magnetic field and the applied voltage, equilibrium will be reached between the applied voltage and the back e.m.f.

The applied voltage is given the symbol V
The back e.m.f. is given the symbol E

In order that current shall flow into the motor in order to produce the necessary force and torque to sustain rotation, E must always be less than V.

$V - E = IR$ where the conductor current $= I$ amperes and the conductor resistance $= R \; \Omega$

If the mechanical load on the motor is increased it will cause it to slow down so reducing the velocity of the conductors through the field. This reduces the back e.m.f. so causing the current to increase. The increase in current provides the extra force necessary to keep the motor turning against the increased load.

Power, force and torque

Figure 3.2 shows a pulley connected to a motor. It is arranged to wind up a rope, the downward force in which is F newtons. The force is provided by a suspended mass. In one revolution of the pulley the mass is raised through $2\pi r$ metres.

$$\text{Work done} = \text{Force} \times \text{distance} = F \times 2\pi r$$

The pulley rotates at n rev/s.

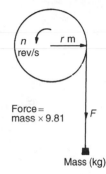

Figure 3.2

Therefore in one second, work done $= F \times 2\pi r \times n$ joules
Now 1 joule per second $= 1$ watt and $F \times r =$ torque T
Hence Power $= 2\pi n T$ W

Fleming's rules

- **For motors.** The relationship between the direction of the force, current and magnetic field has been determined by experiment and a convenient way of remembering this is by the use of Fleming's left hand rule.

   In Figure 3.3 the leFt hand rule relates current and magnetic field to Force and force is produced by an electric motor to drive its load. The ThuMb indicates the direction of the force and hence Motion. The First finger represents the direction of the Field. The seCond finger represents the direction on the Current.

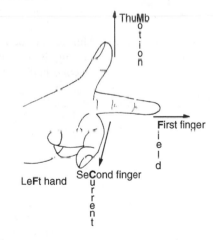

Figure 3.3

- **For generators.** For Generators the riGht hand is used. The thumb and two fingers represent the same quantities as for motors. In a generator the conductor system is driven by an external force and current flows such as to produce a force which opposes that motion. This is clear from Figures 3.4 and 3.5. For the generator, the motion and field are unchanged but the current direction has reversed.

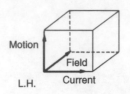

Figure 3.4

Figure 3.5

The action of the commutator

Consider now the very simple generators shown in Figure 3.6. In both cases the magnetic fields are stationary and the coil rotates. Figure 3.6(a) is an alternating current generator and the output is taken from the coil using conducting slip rings on which blocks of carbon called brushes rub. Early machines used wire gauze made up like a sweeping brush and the name is still used. With the coil in the position shown the generated e.m.f. is zero since the coil sides are not cutting the lines of flux. As the coil revolves the voltage increases to a maximum after a rotation of 90° when the coil sides are moving directly across the flux. After a further 90° rotation the voltage will again be zero. As the rotation continues a voltage of reversed polarity will be produced, rising to a maximum and falling to zero as the revolution is completed. The polarity is reversed since the direction of motion of the coil sides is reversed during this period. A coil side which moved from right to left during the first half revolution moves from left to right during the second half revolution.

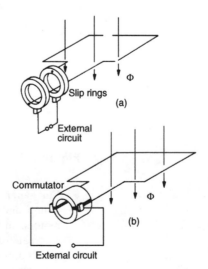

Figure 3.6

Now consider the same coil but this time connected to a two-segment commutator as shown in Figure 3.6(b). This simple commutator is a split copper cylinder, each half fully insulated from its neighbour. As before, carbon brushes are used to connect the external circuit.

The action of the commutator can be understood by looking firstly at Figure 3.7. The brushes are shown inside the commutator for clarity. Assume that there is an external load connected so that current flows in the coil. The left hand coil side connected to commutator segment 1 (hatched) has the current direction shown. This is deduced from Fleming's right hand rule. Current is flowing in the coil towards the left hand brush, which delivers current to the external circuit and therefore has positive polarity. Current returns from the circuit to the right hand brush and flows in the coil away from the commutator segment. (Current direction is shown $\oplus$ which is current flowing away from the viewer whereas a circle with a dot at its centre indicates current flowing towards the viewer.)

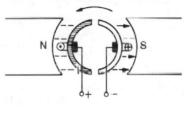

Figure 3.7

In Figure 3.8 we see the coil turned through 90°. The coil sides are outside the field so that the generated e.m.f. is zero. The brushes are shorting out the coil since they are touching both halves of the commutator at the same time. Later, when commutators with very many more segments are considered it must be remembered that the change of connection from one segment to the next must occur when the associated coil sides are out of the magnetic field.

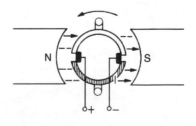

Figure 3.8

In Figure 3.9 the coil has turned through a further 90° and commutator segment 1 is now on the right. The current direction in the coil side connected to it has reversed but since a commutator is being used the left hand brush is now connected to the other side of the coil and so still has the same polarity.

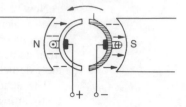

Figure 3.9

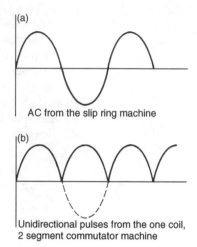

(a)

AC from the slip ring machine

(b)

Unidirectional pulses from the one coil,
2 segment commutator machine

Figure 3.10

The e.m.f. generated in the coil is alternating but by using a mechanical reversing switch or commutator, the current flowing in the external circuit is a series of unidirectional pulses.

The ring-wound armature

In Figure 3.11 the laminated iron core is wound with ten coils each having two turns. The commutator has ten segments, one for each coil. The magnetic flux produced by the poles crosses the air

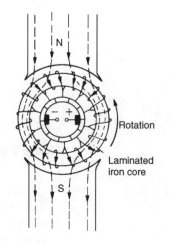

Figure 3.11

gap into the core. Only the outsides of the turns directly under the poles cut the flux and the directions of the induced e.m.f.s shown are again deduced using Fleming's right hand rule. In both the top and bottom halves of the winding this direction is towards the right hand brush which is therefore positive.

With the armature in the position shown each brush is shorting out one coil which has no e.m.f. induced in it since it is not cutting the flux. The flux at this instant is passing through the coil which is the required condition for commutation as already described (see Figure 3.8). Each coil passes through this position in turn as the armature rotates and during this instant carries no current.

There are six conductors cutting the magnetic flux from each pole at any instant in the arrangement shown in Figure 3.11. Representing the e.m.f. induced in each conductor by a cell, each half of the winding has six cells in series. Both halves of the winding are in parallel.

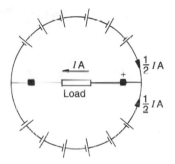

Figure 3.12

When a load is connected each half of the winding carries one half of the load current. As with a single coil, the direction of the e.m.f. induced in individual coils is alternating, being in one direction as the coil passes under the north pole and in the opposite direction as it passes under the south pole. In this machine the air gap between the pole faces and the iron core is of uniform length so that the flux is uniform in the gap.

As a coil passes under a pole from point A to point B in Figure 3.13, the induced voltage is nearly constant. Whilst moving from point B to point C the voltage falls to zero since no flux is being cut. The waveform of the voltage in a single coil is shown in Figure 3.14. This should be compared with that in Figure 3.10 which was for a different coil and field configuration.

There are two disadvantages to the ring winding. These are:

1 Winding the ring is difficult since each turn has to be taken round the core by hand.
2 Only a small part of the winding is active at any one time.

Other windings have been developed to overcome these problems but an understanding of the principles involved in this simple machine is a great help in understanding the following two main types.

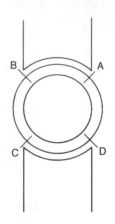

Figure 3.13

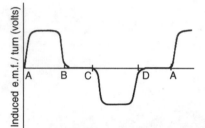

Figure 3.14

The lap winding

Each coil in the lap winding overlaps its neighbours.

Figure 3.15 shows a lap coil comprising two turns. It is coil number 1 and its two sides are labelled 1 and 1' respectively. Its two ends are connected to adjacent segments on a commutator and coil number 1 starts on segment number 1. The other end of the coil is connected to segment number 2.

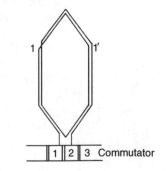

Figure 3.15

Coil number 2 is added and this starts on commutator segment number 2. Only one turn per coil is shown here making the diagram easier to follow. There may in fact be many turns on each coil.

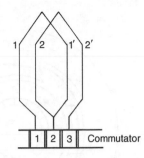

Figure 3.16

The coils are situated in slots in a laminated iron armature and Figure 3.17(a) shows a part-wound armature.

Figure 3.17(b) shows how the windings overlap. The complete winding is built up in this manner until all the slots in the armature contain two coil sides, one on top which starts at the commutator segment which carries its number, 1, 2, 3 etc. and the other at the bottom which is the return coil side numbered 1′, 2′, 3′ etc. The fully wound armature rotates in a magnetic field which is produced electrically. A typical four-pole arrangement together with its commutator construction is shown in Figure 3.18.

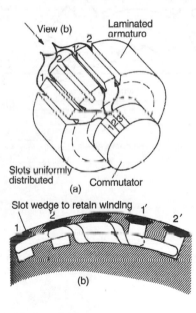

Figure 3.17

Consider a typical section of the lap winding under a north and south pole as shown in Figure 3.19. The view is from the centre of the armature looking outwards through the winding and seeing the pole faces outside the winding. The lines of force from the north pole are straight up out of the paper and those from the south pole down into the paper. Using Fleming's right hand rule the directions of the induced e.m.f.s are as shown. Between the positive and negative brushes there are six conductors connected in series so that the e.m.f. between the brushes is six times that

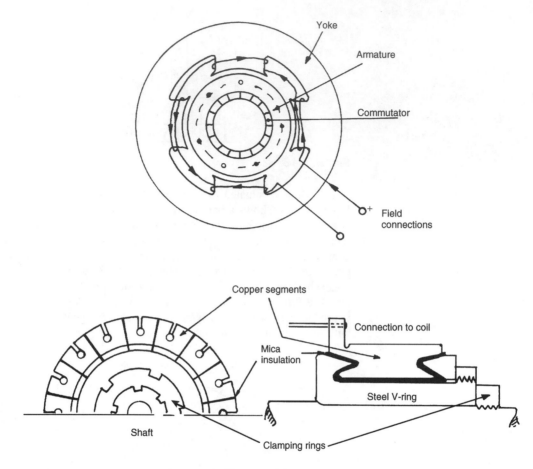

The commutator

Figure 3.18

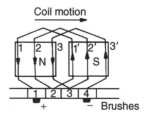

Figure 3.19

induced in a single conductor. The complete lap winding has one brush for each pole and the brushes under like poles are connected together. For a four-pole machine there are four brushes and the arrangement is shown in Figure 3.20.

The winding shown in Figure 3.19 is repeated four times; between brushes A and B, B and C, C and D, and D and A. The full winding on this simple machine therefore comprises 24 conductors (12 turns), conductor 12′ returning to segment 1 under conductor 3 to complete the winding.

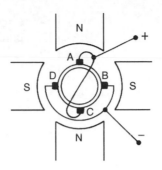

Figure 3.20

There are therefore four parallel paths through the armature: from B to A, B to C, D to A and D to C. Each path comprises six conductors in series.

Example 3.1

A six-pole, DC, lap-wound generator has 36 slots on the armature. Each slot contains eight conductors. When the speed of the machine is 500 rev/min, the induced e.m.f. in each conductor is 5 V.

Calculate: (a) The total number of conductors on the armature.
        (b) The total number of turns on the armature.
        (c) The number of conductors in series in each of the parallel groups.
        (d) The total generated e.m.f.
        (e) The rating of the generator if each conductor can carry 10 A before overheating.

(a) Total number of conductors = number of slots
$$\times \text{ number of conductors in each slot}$$
$$= 36 \times 8$$
$$= \mathbf{288}$$

(b) Two conductors make up one turn.
Number of turns = 288/2 = **144**.

(c) Since there are six poles, there will be six brushes and six parallel ways through the armature.

Number of conductors in series in each group $= \dfrac{288}{6}$
$$= \mathbf{48}$$

(d) Generated e.m.f. = e.m.f. per conductor
$$\times \text{ number of conductors in series}$$
$$= 5 \times 48$$
$$= \mathbf{240\ V}$$

(e) Each conductor can carry 10 A and there are six parallel paths.

$$\text{Maximum current from the armature} = 6 \times 10$$
$$= 60 \text{ A}$$
$$\text{Rating} = 240 \times 60$$
$$= \textbf{14 400 W}$$

The wave winding

Figure 3.21 shows a coil for a wave winding. The ends of the coil are not connected to adjacent segments on the commutator but to segments some distance apart. Again, using a four-pole machine as an example, the winding is shown in part in Figure 3.22.

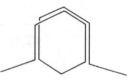

Figure 3.21

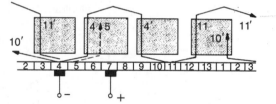

Figure 3.22

Notice that commutator segments 2 and 3 are repeated in the figure at each end so that the commutator may be visualized in its circular form. There is an additional commutator segment, number 13, which causes the winding to progress in 'waves'.

Starting with the negative brush presently on segment 4, follow conductors 4 and 4′ to segment 11, and then conductors 11 and 11′ to segment 5. Four conductors are involved in progressing one segment along the commutator. The process may be continued following four more conductors (not shown) so returning to segment 6. A further four conductors, making 12 in all, allows us to progress to segment 7 upon which the positive brush rests.

With this winding two parallel paths through the armature exist. One of them is as described and the other can be traced from segment 4 once again, but going the other way through conductor 10′ which passes under the pole on the right of the diagram, eventually returning to segment 7.

A wave-wound machine has only two brushes and two parallel paths through the armature irrespective of the number of poles.

A DC generator has four poles and 72 conductors on the armature. The e.m.f. generated per armature conductor at the rated speed is 10 V. The current-carrying capacity of each conductor is 20 A. Calculate the value of the generated e.m.f. and maximum power rating for:

(a) the armature wound as a lap machine
(b) the armature wound as a wave machine.

Example 3.2

A four-pole, wave-wound generator has the following details:

Number of slots on the armature 64
Conductors per slot 15
Induced e.m.f. per conductor at rated speed 1.8 V
Maximum current per conductor 15 A.

Determine: (a) The output voltage.
(b) The maximum current which the machine can safely deliver.

(a) Total number of conductors $= 15 \times 64 = 960$
There are two parallel paths, hence the number of conductors in series per path $= \dfrac{960}{2} = 480$
Total induced e.m.f. $= 1.8 \times 480 = \mathbf{864\ V}$

(b) Maximum current $= 15$ A in each of two parallel paths
$= 15 \times 2$
$= \mathbf{30\ A}$

From these calculations it should be apparent that the lap winding with its potential for many parallel paths (associated with an equal number of magnetic poles) is suitable for higher current, lower voltage applications. The wave machine on the other hand can only ever have two parallel paths and so deliver lower currents but at higher voltages because of the greater number of conductors in series.

The e.m.f. and speed equations

Figure 3.23 shows part of a DC machine. Consider a single conductor on the armature as it moves from position A to position B.
Let ϕ = total flux per pole in webers
n = speed of the armature in revolutions per second
P = number of *pairs* of poles on the yoke. (Pairs are used since it is not possible to have a single pole.)
Z = total number of conductors on the armature
c = number of parallel paths through the armature.
There are $2P$ poles on the machine and at n rev/s the particular conductor will therefore pass, and link with the flux of $2Pn$ poles every second.

$$\text{Time taken to pass one pole} = \frac{1}{2Pn}\ \text{s}$$

This is the time taken for the conductor to move from A to B in Figure 3.23. During this time the flux ϕ from one pole is cut.

$$\text{For a single conductor } e = \frac{\text{flux cut}}{\text{time (s)}}\ \text{volts}$$

Hence, the average e.m.f. induced in the conductor $= \phi \left/ \dfrac{1}{2Pn} \right.$
$= 2Pn\phi$ volts

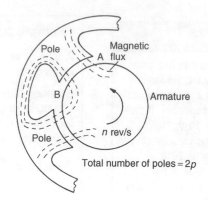

Figure 3.23

On the armature there are Z conductors arranged in c parallel paths so that each group comprises Z/c conductors in series. (See Examples 3.1 and 3.2.)

$$\text{Hence, the total e.m.f. } E = 2Pn\phi\frac{Z}{c} \text{ volts}$$

For the speed equation: transposing gives:

$$n = \frac{E}{2P\phi\dfrac{Z}{c}} = \frac{Ec}{2P\phi Z} \text{ rev/s} \tag{3.1}$$

Since 1 revolution $= 2\pi$ radians

$$\omega = \frac{2\pi Ec}{2P\phi Z} = \frac{\pi Ec}{P\phi Z} \text{ rad/s}$$

A lap-wound DC generator is to have an output voltage of 500 V when running at 26 rev/s. The armature has 28 slots each containing 12 conductors. If a current of 1 A in the field winding produces a flux of 0.08 Wb, determine the required value of field current.

Example 3.3

The wave-wound armature of a six-pole DC generator has 30 slots and in each there are eight conductors. The flux per pole is 0.0174 Wb. Calculate the value of the e.m.f. generated when the speed of the armature is 1200 rev/min.

$$1200 \text{ rev/min} = \frac{1200}{60} = 20 \text{ rev/s}$$

For a wave-wound armature $c = 2$. 6 poles $= 3$ pairs of poles, hence $P = 3$.

$$E = 2 \times 3 \times 20 \times 0.0174 \times \frac{30 \times 8}{2} = \textbf{250.6 V}$$

DC machine connections

- **Shunt.** Figure 3.24 shows the shunt connection. The field winding comprises many turns of fairly fine wire and it is connected in series with a control rheostat directly across the supply. It is therefore in parallel with, or shunting, the armature.

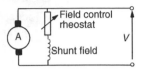

Figure 3.24

The field current is independent of the armature current unless this is large enough to cause a voltage drop within the armature when operating as a generator or in the supply lines when drawing current from a supply to operate as a motor when both armature terminal voltage and field voltage will be affected.

Where it is required that the field current shall be independent of what is happening to the armature, the field circuit may be fed from a separate supply, often derived from an AC source via a rectifier circuit. In this case it is known as a *separately excited* machine.

- **Series.** Figure 3.25 shows the series connection. The field winding comprises a few turns of very heavy gauge wire or copper strip. It is connected in series with the armature and so carries the same current. For this reason it must have a very low resistance or the power loss will be excessive.

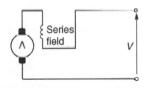

Figure 3.25

- **Compound.** Figures 3.26(a) and (b) show machines with some shunt field and some series field. Both shunt and series coils are wound on the same pole pieces and the series field may either assist or oppose the shunt field. In the former case the machine is said to be *cumulatively wound* and in the latter case *differentially wound*. Figure 3.26(a) shows what is

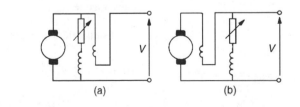

(a)   (b)

Figure 3.26

known as the short shunt connection and Figure 3.26(b) shows a long shunt.

Activity 3.1

Tests on a DC shunt-wound generator.

From the e.m.f. equation for a DC machine we see that e.m.f. is directly proportional to speed and to field flux. These relationships can be verified by carrying out two tests on a separately excited shunt generator.

(a) From the name plate establish the normal rated field current for the machine or if this is not readily apparent, measure the field current drawn from a DC source of value equal to the name-plate output voltage of the machine. Separately excite the machine and use a control rheostat to set the field current at 50 per cent of the normal value. Using an external variable speed driving motor increase the speed of the armature taking readings of the armature terminal voltage and speed for a number of steps up to full-rated speed. Plot a graph of output voltage/speed. Is it linear? Does this verify the voltage/speed relationship in the e.m.f. equation?

(b) Increase the field current in steps of 10 per cent repeating the tests at each step until 100 per cent of the rated field current is reached.

Increases in field current of 10 per cent might be expected to give 10 per cent increases in the voltages recorded for specific speeds. Is this the case?

Any non-linearity may be shown by extracting data to enable you to plot a curve of output voltage/field current for a particular speed or speeds.

Investigate what is meant by the term *magnetic saturation*. Discuss and report on your findings.

The DC motor

Speed/armature current characteristics

When a voltage V is applied to the armature of a motor, a current I_a flows. A force is produced which causes the armature to accelerate and a back e.m.f. E volts is generated. Clearly V must always be greater than E since if it were not so then current would not flow into the armature and no work could be done. The armature resistance $= R_a$ Ω.

Equilibrium is reached when $E = V - I_a R_a$ \hfill (3.2)

Transposing: $I_a R_a = V - E$. Therefore $I_a = \dfrac{V - E}{R_a}$

Now $n = \dfrac{Ec}{2P\phi Z}$ rev/s   (Speed Equation 3.1 above)

For a particular machine P, Z and c are constants so let $\dfrac{2PZ}{c} = k$, a constant.

So that now $\quad n = \dfrac{E}{k\phi}$ $\hspace{3cm}$ (3.3)

From Equation (3.2) we see that any increase in I_a will cause the voltage drop $I_a R_a$ to increase and the voltage E to decrease.

In a shunt-wound motor if the field rheostat is not adjusted, the flux ϕ will remain constant and so from Equation (3.3) we see that since both k and ϕ are constant, E and n are directly proportional.

In a series motor the field winding carries the armature current. The resistance of the armature circuit must include that of the series field since it will cause a voltage drop in addition to that in the armature itself.

Hence $\quad E = V - I_a(R_a + R_{sf})$ $\hspace{2.5cm}$ (3.4)

where $R_{sf} =$ resistance of the series field.

Ignoring the effects of magnetic saturation, the flux set up by the field winding is proportional to the field current and so to the armature current since in the series machine these are the same thing.

Again, using Equation (3.3): $n = \dfrac{E}{k\phi}$

Since increasing the armature current increases the flux, for a constant value of E the effect of increasing the load (and I_a) is to reduce the speed. In fact E falls as the armature current increases (Equation 3.4 above) so that there is a further reduction in speed due to this.

Example 3.4

A DC shunt motor has an armature circuit resistance of 0.5 Ω and a shunt field resistance of 240 Ω. It is connected to a 240 V supply. On no load the input current is 2 A and the speed 1500 rev/min.

When fully loaded the input current is 8 A. Calculate the value of back e.m.f. generated and the speed of the motor on full load.

1500 rev/min = 25 rev/s

$I_a = I - I_f$ (from Figure 3.27). Therefore
$I_a = 2 - 1 = 1$ A on no load
$E_1 = V - I_a R_a$ (using suffix 1 to indicate original no-load
$\hspace{1.5cm}$ conditions)
$\hspace{0.8cm} = 240 - (1 \times 0.5)$
$\hspace{0.8cm} = 239.5$ V

$\qquad n_1 = \dfrac{E_1}{k\phi}$ $\quad$ Transposing: $k\phi = \dfrac{239.5}{25} = 9.58$

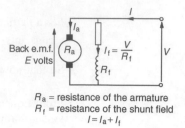

R_a = resistance of the armature
R_f = resistance of the shunt field
$I = I_a + I_f$

Figure 3.27

When the total current = 8 A, $I_a = 7$ A. (The field current is unchanged at 1 A)

$$E_2 = 240 - (7 \times 0.5) \text{ (Suffix 2 indicates the new, loaded condition)}$$

$$= \textbf{236.5 V}$$

$$n_2 = \frac{E_2}{k\phi} = \frac{236.5}{9.58} = \textbf{24.68 rev/s}$$

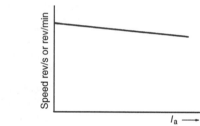

Figure 3.28

Figure 3.28 shows a typical speed/armature current characteristic for a shunt motor.

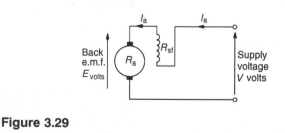

Figure 3.29

Activity 3.2

Carry out a load test on a shunt-wound DC motor.
   DC instruments are required to measure supply voltage and armature current. (Total current I [Figure 3.27] may be measured and a calculated value of field current I_f

deducted if this is more convenient.) The motor is started on no load and then loaded using any form of dynamometer. The speed is measured using either a contact tachometer or permanently installed device.

Record readings of armature current, supply voltage, torque developed and speed for a range of loads up to full load for the machine.

Plot graphs of (a) speed/I_a and (b) torque/I_a. Comment on their forms.

Calculate the efficiency of the motor at each stage of loading.

Output power $= 2\pi nT$ watts: Input power $= VI$ watts. Plot a graph, efficiency/output power.

Comment on the shape of the efficiency/output curve.

What is the effect of changing the field current? Carry out a further test with a reduced field current such that the no-load speed is increased by, say, 10 per cent.

Relate the results to the theory.

Example 3.5

A series motor has an armature resistance of 0.15 Ω and a series field resistance of 0.25 Ω. It is connected to a 250 V supply and at a particular load runs at 30 rev/s when drawing 10 A from the supply. Calculate the speed of the motor when the load is such that the armature current is increased to 20 A.

$$E_1 = V - I_a(R_a + R_{sf})$$

$$= 250 - 10(0.15 + 0.25)$$

$$= 246 \text{ V}$$

As in Example 3.4,

$$k\phi_1 = \frac{E_1}{n_1} = \frac{246}{30} = 8.2$$

When $I_a = 20$ A, $E_2 = 250 - 20(0.15 + 0.25) = 242$ V

Now, since the armature current has doubled, the flux has doubled (ignoring saturation effects).

Therefore $k\phi_2 = 2 \times 8.2 = 16.4$

$$n_2 = \frac{E_2}{k\phi_2} = \frac{242}{16.4} = \textbf{14.93 rev/s}$$

Thus, as the armature current increases the speed falls. Figure 3.30 shows a typical speed/armature current characteristic for a series motor. Notice that at low armature currents the speed is high since the flux is low and for this reason the series motor must always be connected to a load which will limit its top speed. On no load the speed would be sufficiently high to create disruptive centrifugal forces and the commutator segments and windings would be thrown outwards.

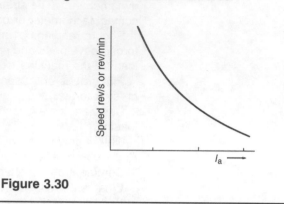

Figure 3.30

Activity 3.3

Carry out a load test on a DC series motor.

The instrumentation and loading requirements are as in Activity 3.2 for the shunt machine. There will of course be no separate field current to be measured.

Making sure that there is always some load on the machine to limit the top speed, vary the load whilst measuring the electrical input, the torque developed and the speed for a range of loads up to that giving the maximum permitted armature current.

Record your readings. Plot curves showing the relationships between (a) armature current and torque and (b) armature current and speed.

Discuss these shapes and relate them to the theory involved.

If there is provision for changing the number of turns on the series winding or for bypassing part of the armature current round the field winding, repeat the tests with a slightly smaller number of turns in use/less current in the field winding. Discuss the changes in characteristic which result.

Suggest suitable uses for the series motor. Research some applications which are most suited to the very high torque which is produced at low speed. What measures are taken in practice to ensure that the motor can never be completely unloaded?

(Investigate electric vehicle and railway traction.)

A level characteristic may be produced by compounding. With the shunt motor the speed falls as the load is increased. By adding a small differentially wound series winding to the main poles the total flux is reduced as the load current increases. Since speed is inversely proportional to flux, this causes an increase in speed compensating for the reduction due to the resistance of the armature.

Cumulative windings may be used to give a motor characteristic between that of the series and shunt motors. The speed will fall as the load is increased but the motor will have a safe maximum speed on no load due to the constant shunt field.

The torque equation

From previous work we have, for a shunt motor:

$$E = V - I_a R_a$$

Multiply throughout by I_a

$$EI_a = VI_a - I_a^2 R_a \quad \text{(all watts)}$$

$$VI_a = \text{total power input to the armature}$$
$$\text{(Supply voltage} \times \text{armature current)}$$

$$I_a^2 R_a = \text{power loss in the armature due to heating}$$

Hence EI_a must be the power available at the armature shaft to produce torque. (Note that by using $(R_a + R_{sf})$ throughout, the proof is valid for a series motor.)

Now

$$E = \frac{2P\phi Zn}{c} \text{ volts} \quad \text{and power} = 2\pi nT \text{ watts}$$

Therefore

$$EI_a = \frac{2P\phi Zn}{c} I_a = 2\pi nT$$

Transposing:

$$T = \frac{2P\phi Zn}{c2\pi n} I_a = \frac{P\phi Z}{c\pi} I_a \text{ newton metres}$$

As before, let $\dfrac{2PZ}{c} = k$, a constant for the particular machine

When

$$T = \frac{k}{2\pi} \phi I_a \text{ Nm} \quad \text{where } T = \text{ gross torque.}$$

The gross torque developed is proportional to the flux and to the armature current. Not all this torque is available to drive an external load, however, since there will be friction in the motor bearings and between the armature surface and the air in the casing. The armature may be driving a fan which forces air over the windings to keep them cool and a torque is required to drive this. Finally, since the iron core of the armature is being driven through a magnetic field there will be hysteresis and eddy current losses.

Test your knowledge 3.4

A four-pole lap-wound motor is connected to a 400 V DC supply. The armature has 600 conductors and a resistance of 0.7 Ω. The total flux per pole is 0.08 webers. Calculate:

(a) the speed of the motor,
(b) the motor torque developed when the armature current = 25 A
(c) the net output torque if the armature loss torques total 10 Nm.

Activity 3.4

We know that torque is proportional to magnetic flux and to armature current in a DC motor. Flux varies with field current so that torque is a function of field current and armature current.

Investigate this relationship.

With a DC shunt motor set up as for Activity 3.2 load the machine to a suitable torque, say one half of the full-load value. Record the values of field current, armature current and speed. Using the field control rheostat, very gently make a small change in the field current whilst adjusting the dynamometer as necessary to maintain the same value of torque. Record the new values of field and armature currents together with the speed. Repeat the process for one more change, if practicable, this time changing the field current in the opposite sense to the previous test. Keep within the safe parameters for the machine; make only small changes since decreasing the field too much can result in an unstable condition. To prevent the dynamometer from overheating it may be necessary to off-load the machine between readings, then re-establish the new conditions.

Discuss your results and report on how far they accord with the theory. What effects do the speed changes and magnetic saturation have on the results?

Torque/armature current characteristics

In the shunt motor for a constant field rheostat setting, the flux ϕ is constant.

$$\text{Now } T = \frac{k}{2\pi}\phi I_a \text{ Nm}$$

Therefore I_a is the only variable and the gross torque is proportional to the armature current. This is shown in Figure 3.31.

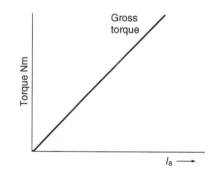

Figure 3.31

In the series motor the field is produced by the armature current. $\phi = k_1 I_a$ up to saturation where k_1 is the constant of proportionality between flux and current in webers per ampere. Substituting in the torque equation:

$$T = \frac{k}{2\pi}k_1 I_a \times I_a = \frac{k}{2\pi}k_1 I_a^2$$

Which is the equation of a parabola $[y = mx^2]$. This is shown in Figure 3.32.

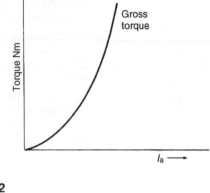

Figure 3.32

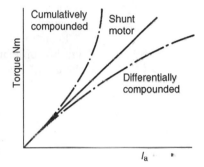

Figure 3.33

Differential compounding will cause the torque produced to be less than that of a shunt motor for a given value of armature current whilst cumulative compounding will increase the available torque.

Motor applications

Shunt motors are used where virtually constant speed is required on drives such as machine tools, fans and conveyor systems.

Series motors produce very high torques at low speeds so that they are suitable for starting very heavy loads. The main uses are in traction where they are extensively used in trains, electric buses, trams and milk delivery vehicles. They are permanently connected to their loads and so cannot ever run with zero applied torque.

Compound wound motors are used where constant speed is required (differential compounding) for coal feeders on boilers and oil pumps or where high torque is required on starting and safe top speeds are required (cumulative compounding) for conveyor systems, hoists and cranes.

Speed control

A measure of speed control may be achieved by changing the field current of the motor. Since the induced voltage E is proportional to flux and speed, any changes in magnetic flux will result in a change in speed. The lower the flux, the higher the speed. However, torque is proportional to flux and armature current so that if the flux is reduced the armature current must increase if the load torque remains constant as the speed changes. This is often the limiting factor since large armature currents will cause overheating. In addition, a large armature current creates a large armature magnetic field which interacts with the main field which can cause severe problems. (See Armature reaction.) Usual values of speed obtainable by field weakening on a standard motor will be from normal up to something less than twice normal although a wider speed range is achievable using specially designed motors. (See Activity 3.4.)

Since steel saturates magnetically at less than 2 tesla, increasing the field current and hence field strength as a method of speed reduction has very limited application. Speed reduction must be achieved by reducing the voltage on the armature. If the field and torque remain constant, the armature current and volt drop will remain constant. Since $E = V - I_aR_a$, as the applied voltage V is reduced so the speed will fall in proportion to satisfy the e.m.f. equation.

Figure 3.34(b) shows the characteristic of a shunt motor employing two different field currents, field variation being achieved by the use of a control rheostat (Figure 3.24).

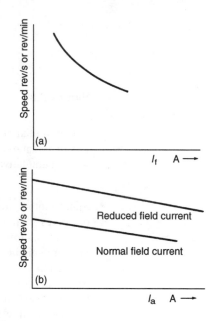

Figure 3.34

In series motors a reduction in flux may be achieved by either bypassing part of the armature current round the field winding or by having a tapped field winding which allows a variable number of turns to be used. These two methods are shown in Figure 3.35.

Reductions in armature voltage can be obtained by using an additional resistor in the armature circuit but this is extremely wasteful of power since the whole of the armature current flowing in this resistor produces a considerable amount of heat. If the speed is reduced to one half normal by this method then one half of the input power is wasted.

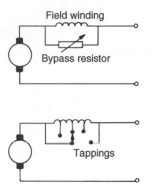

Figure 3.35

Where large motors are involved controlled rectifiers or a DC generator may be used to provide the varying voltage. Silicon controlled rectifiers are often used when the mean input voltage can be controlled from zero to its full-rated value at very high efficiency.

Where a DC generator is used, this is often driven at constant speed using an AC motor. The output voltage from the generator may be varied from zero to full-rated value by increasing its field current. By reversing the field connections, the polarity of the output voltage is reversed. A motor supplied from this generator may be made to run at any speed from a crawl to its full-rated value in either direction. Such a system is named after Ward-Leonard, the developer.

Activity 3.5

Investigate the speed/applied voltage relationship in a shunt motor.

Use either of the methods described in the above paragraphs to obtain a variable DC voltage. Use this variable voltage to feed the armature of a DC shunt motor. Its field must be supplied from a separate DC supply at a fixed value appropriate to the normal full speed value of the motor. The motor runs light: it is not loaded using a dynamometer. Starting with zero input voltage, gradually increase the voltage applied to the armature in steps, measuring the speed of the shaft at each step. Record your readings and plot a graph of speed/armature applied voltage for the set field.

Repeat for another value of field current. Why, when using this method of speed control, would the maximum possible value of field current normally be used?

Discuss your results in light of the theory involved.

Why is this system of speed control much more efficient when a controlled rectifier is used than when a motor-driven generator is employed to obtain the variable-voltage supply? Are there any other advantages to using the rectifier system?

DC motor starters

Since $E = \dfrac{2P\phi nZ}{c}$ volts

it follows that when a motor is at rest, since $n = $ zero, whatever the value of flux, the back e.m.f. is zero.

But $E = V - I_a R_a$. Therefore, at rest $0 = V - I_a R_a$ or transposing: $I_a R_a = V$

$$I_a = \frac{V}{R_a} \text{ amperes}$$

The resistance of the armature is kept as small as possible so as to keep the losses in the armature to a minimum. It follows that the armature current under these conditions will be extremely large. It could in fact be large enough to cause severe damage to the armature by heating the conductors and commutator. Also since there are mechanical forces between current-carrying conductors, with large currents sufficient force may be developed to disrupt the winding.

Measures must be taken to prevent these excessive currents flowing. Where the motor is supplied from direct mains this may be achieved either by the addition of resistance to the armature circuit or by electronic means employing a 'chopper' circuit.

Where the motor is fed from an alternating supply, controlled rectifiers as already mentioned under Speed control will be used to vary the level of input voltage and hence the starting current.

The addition of resistance to the armature circuit

In Figure 3.36 the resistance R ohms is included in the armature circuit so that at standstill

$$0 = V - I_a(R_a + R) \quad \text{or transposing} \quad I_a = \frac{V}{(R_a + R)}$$

The added resistance is made large enough to limit the starting current to between 1.5 and 2 times the normal full-load current of

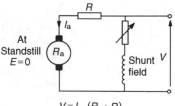

$V = I_a \ (R_a + R)$

Figure 3.36

the motor. As the motor accelerates the back e.m.f. rises from zero so that

$$I_a = \frac{V - E}{(R_a + R)}$$

The value of the added resistance can now be decreased somewhat to allow the armature current to rise to its original value once more. Further acceleration and an increase in the back e.m.f. will take place. Reductions in additional resistance are made at intervals as the speed of the motor builds up. Finally when the armature resistance alone remains

$$I_a = \frac{V - E}{R_a}$$

During the starting process it is essential to have the maximum possible field strength in the motor since both torque and back e.m.f. are proportional to the field flux. The additional resistance must therefore only affect the armature circuit whilst the field winding is connected directly across the supply.

The manually operated face-plate starter

Figure 3.37 shows the armature circuit only of a DC shunt motor with a suitable starter. The handle is moved manually from the off position to make contact with the first resistance stud. This puts resistances R_1, R_2, R_3 and R_4 in series with the armature.

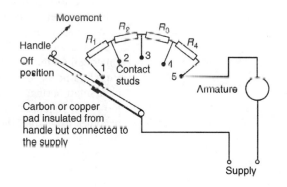

Figure 3.37

As the motor speed rises from zero to a few revolutions per second the handle is moved so that the supply is fed to contact number 2. Resistances R_2, R_3 and R_4 are now in series with the armature. As the speed builds up the handle is moved successively across the studs ending up on number 5 when the armature is connected directly to the supply. As stated previously, the field current must be a maximum during this period and a method of achieving this is shown in Figure 3.38.

The starting handle being moved on to the first stud puts all the resistances in series with the armature. The current from the supply positive terminal flows through O, the overcurrent trip, to

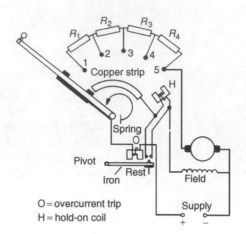

Figure 3.38

the bottom of the starting handle along which there is a conducting strip to two contact pads. One of these feeds current to the series resistors whilst the other makes contact with a copper strip which is connected to the field winding. On the way to the field winding the current flows through H, the hold-on coil. The field winding is therefore subjected to the full supply voltage (less a very small volt drop in coil H). As the starting handle progresses to the running position the field remains unchanged.

When the handle reaches stud 5 it comes into contact with the faces of the electromagnet H. Either the whole handle or a small pad fixed to it is made of a magnetic material, so provided that a suitable current is flowing in the field winding, the handle will be held in the running position.

All the current being supplied to the motor flows through the small coil on the horseshoe-shaped piece of iron of the overcurrent trip O. A beam beneath the coil is held on the rest stop by gravity or a small spring. Normally the magnetic effect of the current is too small to attract the beam upwards. If, however, the current becomes excessive, the beam is pulled upwards and the two electrical contacts to the right are made which causes the field current to be bypassed round the hold on coil. This becomes de-energized and the starting handle is released. A spring causes it to fly back to the start position.

If at any time the supply to the motor is lost, the coil H will be de-energized and the handle will return to the start position. This prevents the motor from starting up unexpectedly when the supply is restored with possibly disastrous results both to the motor and to the operator who may have a hand in the driven equipment.

Automatic contactor-type starter

The manually operated starter suffer from a number of disadvantages amongst which are (1) there will always be some burning of the contacts, especially in the larger sizes, as the operator moves the handle across the face of the starter and (2) there is no real control over the speed at which the handle is moved (apart from possibly a separate external overload trip which would cut off the supply if the input current reached a very high value). Better types of starter involve the use of contactors as shown in Figure 3.39.

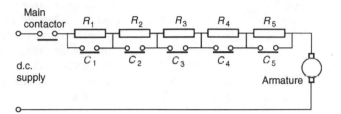

Figure 3.39

The motor is started by closing the main contactor. All the resistance is in circuit, as in the case of the manual starter. The motor speeds up when contactor C_1 is closed. This shorts out R_1 allowing the motor to gain more speed. Progressively the contactors are closed until finally, by closing C_5 the armature is connected directly across the supply. The field has to be connected to the full voltage during the run-up period, but this is not shown in the figure.

The relays may be closed by using relays actuated:

1 by the build-up of armature-generated voltage E. For example, with a motor being supplied from 400 V mains, when E reaches 50 V, C_1 closes, when E reaches 100 V, C_2 closes etc.
2 by the swings in armature current. When the motor is started the initial input current will be of the order of twice the full-load value. As the speed increases, the current value will fall. When it reaches 1.5 times the full-load value, C_1 closes, causing the current to rise and the speed to increase. When the current falls again to 1.5 times its full load value, C_2 closes and so on until all the contacts are closed.
3 on a time basis. Several run-ups are performed manually and the times taken to reach certain key speeds noted. These values are fed into a timing circuit which will close the contactors after these set periods during each run up.

Electronic means

There is a considerable power loss when using the resistance type of starting. These losses may be avoided and often a much smoother run-up achieved using electronic starters. We will consider starting when employing (a) a direct supply and (b) an alternating supply.

(a) *Direct supply.* The electronic chopper switches the mains direct supply across the armature for a very short period of time, typically a few milliseconds. The current begins to rise towards a very high value but before this is achieved the supply is disconnected. After a further very short period the supply is reconnected and again disconnected. The switching takes place many times per second and over a number of seconds. The motor will be gathering speed all the while and the 'on' periods will be made progressively longer. This is shown in condensed form in Figure 3.40. There will be hundreds, if not thousands, of switching operations before the chopper circuit ceases to operate and the armature is connected directly to the supply. It is possible to hear the chopper at work on vehicles such as milk floats as a note that varies as the vehicle gathers speed. Some electric trains emit this varying note indicating the use of a chopper whilst with others the 'tick, tick' noise of closing contactors indicates the control used.

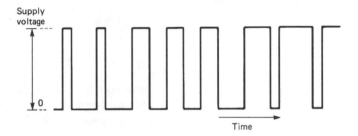

Figure 3.40

(b) *Alternating supply.* A controlled rectifier employing thyristors is used to vary the voltage applied to the armature. An electronic circuit is used to cause the rectifier to become conducting at any particular point in the alternating cycle. Delaying conduction for most of the cycle means that the average voltage supplied to the armature is low. Reducing the delay means that a higher average voltage is delivered.

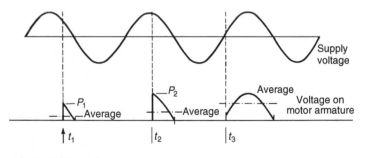

Figure 3.41

In Figure 3.41 a single-phase half-wave rectifier is considered for ease of explanation. During the positive half cycle the rectifier is switched on at t_1. The voltage delivered to the motor has peak value p_1. The average voltage over the half cycle is low so that the input current to the motor is not excessive. After many half cycles, during which the motor is speeding up, the delay on the rectifier is changed to t_2, the peak voltage applied to the armature rises to p_2 and the average voltage is somewhat higher. After a suitable interval, the delay is further reduced and the figure shows the instant just before the complete wave is used, t_3. The average voltage is now much higher. Generally a full-wave rectifier giving twice the number of pulses or a three-phase bridge rectifier giving six times the number of pulses illustrated will be used.

Very fast run-up times are achieved using electronic means since the switching times are controlled by the maximum allowable motor current, this being maintained during the run-up by automatically adjusting the pulse rate with DC choppers or phase angle with thyristors.

Test your knowledge 3.5

A resistance starter as shown in Figure 3.38 is used to limit the current input to the armature of a DC shunt motor to 50 A when connected to a 200 V supply. The armature resistance of the motor is 0.4 Ω. Calculate:

(a) the value of the required additional resistance in the starter at starting
(b) the value of armature current when the speed of the motor has risen so that the back e.m.f. = 50 V, all the resistance remaining connected
(c) the new value of armature current at the speed in (b) above at the instant the starting resistance is decreased by 0.7 Ω.

Activity 3.6

Investigate the action of a DC motor resistance starter. Isolate the resistance-type starter from the electrical supply and measure the resistance of each stage of the starter. Why are they not all of equal value? Which of the stages has the greatest resistance? Why is this? Is there any mathematical relationship between the values of successive stages? (Such a relationship may be only approximate.)

Start a DC motor using the starter. At each stage measure the voltage across the armature, the speed of rotation and the minimum current just before switching out a stage of the starter. Observe how the current varies during the switching stages. (Maximum currents may not be recordable due to their transient nature and the inertia of the meter.) Relate your observations to the theory involved. Are the voltages and speeds at each stage related? Comment on the values of minimum current achieved between each switching operation.

The DC generator

Shunt wound

If a DC machine connected to a supply and running as a motor has the speed of its armature raised above that at which it operates as a motor by driving it with an external engine whilst keeping the field constant, the value of the generated e.m.f. will increase since this is proportional to speed.

As E becomes greater than the supply voltage a current will flow from the machine which has now become a generator, receiving power from the driving engine and delivering power to the electrical system. The difference between E and V is still $I_a R_a$.

$$E = V + I_a R_a$$

For a motor $E = V - I_a R_a$ and the sign change for the generating condition indicates a change in the direction of the current. In the self-excited generator in Figure 3.42 part of the armature output is used to excite its own field so that $I = I_a - I_f$. An increase in load current causes $I_a R_a$ to increase and the terminal voltage V will fall.

The field current $= V/R_f$ so that a reduction in terminal voltage results in a reduction in field current and hence field flux.

Since $E = k\phi n$ volts, the reduction in flux causes a reduction in generated e.m.f. The terminal voltage of a self-excited generator therefore falls quite rapidly as the load current increases due to the combined effects. A typical characteristic is shown in Figure 3.43.

When driven at constant speed with no load connected, the relationship between field current and generated e.m.f. may be determined using the circuit shown in Figure 3.42 with the addition of a voltmeter to measure output voltage and an ammeter in the field circuit. Due to magnetic saturation the output voltage is not proportional to the field current. With zero field current there is a small output voltage which is produced by the residual magnetism in the pole pieces without which the generator cannot commence generation. In a new machine this is created using an external power source. Figure 3.44 shows the open-circuit characteristics.

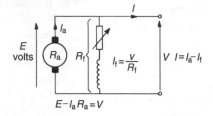

Figure 3.42

Figure 3.43

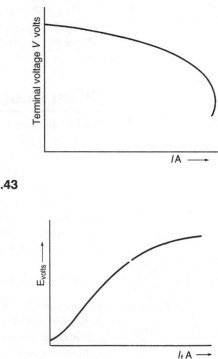

Figure 3.44

Activity 3.7

Conduct a load test on a DC shunt-wound generator. Add a resistive loading bank to a DC shunt generator as shown in Figure 3.42. Add instrumentation to allow terminal voltage, load and field currents to be measured. Run the generator at a constant speed as indicated on its name plate. Increase the load on the generator by changing the settings on the loading bank. Record the output voltage, load and field currents for each load change. It may be possible to increase the load to such a point that the terminal voltage collapses, but do not severely overload the machine to achieve this. Plot the load characteristic of the generator. Explain the theory involved and relate your results to this theory. What change to the characteristic would be apparent if the machine were separately excited?

Series wound

This is rarely if ever used since with no load on the machine there is no current in the field winding and the output voltage is near zero. When a current flows a flux is produced and the output voltage rises. The terminal voltage is therefore a function of the load current.

Compound wound

Cumulative compounding is used to overcome the drooping characteristic shown in Figure 3.43 for the shunt machine. As load current flows in the series winding a flux is produced which is added to that of the shunt field. This increase in total flux produces a corresponding increase in generated e.m.f. The output voltage can therefore be held constant or made to increase slightly to compensate for the voltage drop in the connecting cables to the external load.

Differential compounding is used to cause the terminal voltage to collapse as load current flows. Load current increases the series field which opposes the shunt field so causing a reduction in flux and generated e.m.f. This is used in electric arc welding sets where approximately 110 V is required to strike the arc but only 20 V to maintain the arc whilst welding.

Armature reaction

Further work on communication

Figure 3.45(a) shows one coil of the lap winding originally shown in Figure 3.19. The negative brush is resting on segment 4 of the commutator. As viewed the current direction in the coil 3,3′ is anticlockwise. In Figure 3.45(b) the conductors have moved a distance equivalent to one half of a commutator segment to the right. Both coil sides are now between poles and there is no induced

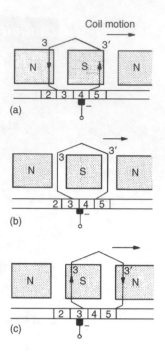

Figure 3.45

e.m.f. in the coil. The brush now shorts out the coil since it spans segments 3 and 4.

In Figure 3.45(c) the coil has again moved to the right a further half segment of the commutator and both coil sides are under poles. The currents are now as shown, the direction being clockwise as viewed.

This means that in the time taken for the armature to move from position (a) to position (c) the current in the coil must reverse in direction. If the coil has a large inductance which is likely since it is embedded in a large mass of iron, and again if the current is large, there will be considerable induced voltage in the coil during the period of reversal. This is known as reactance voltage.

The reactance voltage $e = L\dfrac{\Delta i}{\Delta t}$ volts, where L = inductance of the circuit in henrys and $\Delta i/\Delta t$ = rate of change of current in amperes per second.

Severe sparking will occur at the surface of the commutator if the current reversal, or commutation, is not complete by the time that the commutator segment moves from under the brush.

During commutation the sides of the coil must be outside the magnetic field so that the generated e.m.f. is zero and to achieve this the brushes are moved round the commutator until they are located on what is termed the neutral axis.

Armature reaction

In Figure 3.46(a) a simple armature is shown revolving between two magnetic poles. The current directions are as shown. The coil A, A′ which lies horizontally has no e.m.f. generated in it and the current in it is undergoing commutation. The coil B, B′ which is in a vertical plane at this instant is shown separately in Figure 3.46(b). The current in conductor B flows away from the commutator end along the length of the armature, crosses over at the back and then flows back in conductor B′. This current produces a magnetic north pole on the left hand side of the armature. The other two current-carrying coils will add to the effect. The strength of this cross-magnetic field will depend on the number of turns and the value of the armature current.

In Figure 3.47 two ways of illustrating the effect of armature or cross flux are shown. In Figure 3.47(a) the distorting effect of the current in the two sides of coil B, B′ is shown. In Figure 3.47(b) the vector addition of the armature and main fluxes can be seen. The resultant flux has been displaced by $\theta°$ from the vertical. The effect of the armature flux on the main flux is called armature reaction. The term is not to be confused with reactance voltage which is quite different.

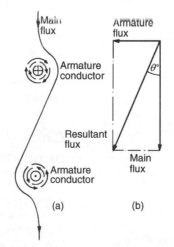

Figure 3.46

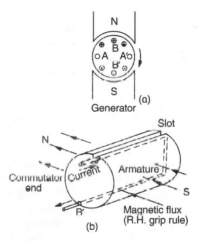

Figure 3.47

In a multi-pole machine where the space between the poles is much smaller than in the two-pole machine this twisting of the field can bring into the field conductors which were out of it and in which commutation is taking place. Severe sparking results. The position for the brushes to be situated in order to obtain satisfactory commutation has now moved $\theta°$ in a clockwise direction from the former position. The new position is known as the magnetic neutral axis and its position is a function of the armature current since the amount of field distortion is dependent on this current. In some older machines brush-moving gear was provided so that an operator could adjust the position of the brushes as the load changed. This is not very convenient and other methods are employed in general to achieve satisfactory commutation under all load conditions.

Compensating windings and interpoles

One method of overcoming the effect of armature reaction is to fit compensating windings to the machine. They carry the armature current and so are series connected. They are arranged to create magnetic poles which oppose those set up by the armature currents so that at any instant the resultant cross field is zero. This type of winding is very expensive and is only used on some special machines.

A more usual alternative is the use of interpoles. These are very thin poles carrying armature current and are situated between the main poles. They are wound to give a polarity which anticipates the action of the next main pole. They assist the reversal of the current in the coil undergoing commutation. The interpole not only neutralizes the effect of armature reaction local to the conductors undergoing commutation but overcomes the effects of reactance voltage.

When considering the action of the interpoles it must be remembered that these always act in the generating mode whether they are fitted to motors or generators. Fleming's right hand rule always applies to them. They assist in inducing a voltage of the correct polarity before the conductor actually moves under the next main pole. In Figure 3.48(a) the simple generator rotating in a clockwise direction has the armature current directions as shown. As the conductor on the right moves from a position X to X′ the current must be established out of the paper as viewed. In order to assist this to occur the interpole field must be as shown. Figure 3.48(b) shows the interpole polarities for a motor with the same rotation.

Changes in characteristics due to armature reaction

Armature reaction causes the magnetic field in a DC machine to be twisted round which concentrates the flux in one side of each pole face so that this side may be driven into saturation as a result. Since the top limit to flux density is in the region 1.6 to 1.8 tesla this may result in a reduction in total flux. The effect is demonstrated in Example 3.6.

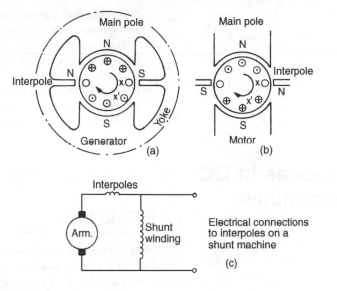

Figure 3.48

Example 3.6

The normal flux density of the poles in a DC generator on no load is 1.1 T. Each pole has a cross-sectional area of 0.05 m². The effect of armature reaction at a particular load is to concentrate all the flux into one half of the pole face area. The saturation flux density is 1.8 T. Calculate the new value of flux per pole and the per unit reduction in total flux at this load.

Total flux per pole = flux density × area of the pole face

$$= 1.1 \times 0.05$$

$$= 0.055 \text{ Wb}$$

Concentrating all this flux into one half of the pole face could be expected to give a flux density of

$$\frac{0.055}{0.05/2} = 2.2 \text{ T}$$

This is above saturation. The flux density cannot rise above 1.8 T. At 1.8 T and an area of 0.05/2 (0.025) m² the flux per pole = 1.8 × 0.025 = 0.045 Wb.
This is a reduction of 0.055 − 0.045 = 0.01 Wb

$$= \frac{0.01}{0.055} = 0.19 \text{ p.u. reduction}$$

In practice the case is never so clear cut; although one side may well be in saturation, the other side will be still carrying some flux. The reduction will not therefore be as large as above but the example serves to demonstrate the principle.

The reduction in flux will cause a further reduction in output voltage of DC generators in addition to those already discussed.

In the case of motors there will be a speed increase as the flux is reduced and in addition, if the load remains constant, the armature current will increase. These effects can be explained by a consideration of the e.m.f. and torque equations.

Losses in DC machines

(See Figures 3.27, 3.29 and 3.42 for connections and symbols.)

Field loss

In the shunt machine the field loss $= VI_f$ or $I_f^2 R_f$ watts.

In the series machine, since the field winding carries the armature current, field loss $= I_a^2 R_{sf}$ watts.

The field loss is a function of the number of ampere turns necessary to magnetize the pole pieces and so is dependent on the grade of steel used.

Armature loss

(a) $I_a^2 R_a$ watts loss in the coils of the armature. For a particular load the armature current is constant so that this loss is directly proportional to the resistance of the winding.

(b) Iron losses. Since the armature is subjected to alternating magnetization as it passes under north and south poles, there will be hysteresis and eddy current losses in the steel. The hysteresis loss is a function of: (1) the maximum flux density, (2) the grade of steel used and (3) the speed of rotation of the armature. Eddy current losses are a function of: (1) the maximum flux density, (2) the thickness of the laminations used, (3) the resistivity of the steel and (4) the speed of rotation.

Commutator losses

(a) Resistance loss. There is resistance between the brush surface and commutator and the loss depends on the grade and quality of the brush used. The volt drop tends to be constant at about 1 V per brush so that this loss is approximately (2 volts × total armature current) watts, irrespective of the type of armature winding.

(b) Friction loss. There is rubbing friction between brushes and commutator surface and the power loss depends on the coefficient of friction and the rubbing speed. The rubbing speed is a function of the diameter of the commutator and the speed of rotation.

Bearing friction

The armature is supported in bearings and these may be of the ball, roller or sleeve types according to motor size and application. The frictional loss is a function of the speed of rotation and the type of bearing.

Windage

There will be friction between the surface of the armature and the air in the casing. In addition power will usually be required to drive the cooling fan. The power loss involved depends on the type and size of the fan and on the running speed.

The power flow diagram

For a motor

Note: The term 'Armature circuit' includes any series field windings and interpoles present.

Starting from the top of Figure 3.49 we have the total input to the motor. From this is subtracted the shunt field loss if applicable. VI_a watts is supplied to the armature circuit where the losses are $I_a^2 \times$ (total resistance of the armature circuit) watts plus the brush resistance loss. The power to create the gross torque is EI_a watts. The gross torque drives the armature against brush and bearing friction, windage and the retarding torque due to hysteresis and eddy currents in the armature. After all the losses have been provided we have the useful output which drives the connected mechanical load.

For a generator

Starting at the bottom of Figure 3.49 we have the total input which is being provided by an engine of some form. The input must provide all the mechanical losses leaving EI_a watts from which

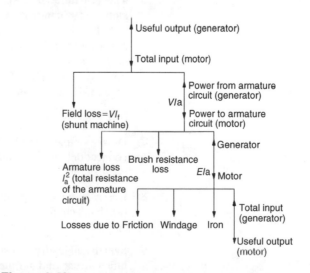

Figure 3.49

the armature circuit losses must be subtracted. This leaves VI_a watts as the output from the armature circuit. Finally in the shunt machine its own field loss must be deducted leaving the useful output at the top.

In both the motor and the generator case the input is greater than the output and the ratio output/input is the efficiency of the machine. Typically this will be between 0.6 and 0.85 according to the rating of the machine.

Problems on DC machines

1   What is the function of a commutator in a DC machine?

2   What is the basic difference between a lap and a wave winding on a DC armature?

3   Which type of winding would you select for: (a) high current, low voltage application, (b) higher voltage, low current application?

4   Calculate the value of the generated e.m.f. delivered by the following six-pole DC generators.

(a)   Lap wound: $\phi = 20$ mWb/pole
number of slots on the armature $= 86$
conductors per slot $= 5$, speed $= 15$ rev/s
(b)   Wave wound: $\phi = 25$ mWb, number of slots $= 24$
conductors per slot $= 10$, speed $= 20$ rev/s.

[(a) 129 V, (b) 360 V]

5   A DC machine has an armature resistance of 0.5 Ω. It is connected to 500 V mains. When drawing 20 A from the supply it runs at 25 rev/s. At what speed must it run as a generator in order to deliver 20 A to the electrical system? The flux remains constant.

[26 rev/s ($E = 510$ V, $k\phi = 19.6$)]

6   The armature of a six-pole, lap-wound DC motor has 54 slots and 8 conductors in each slot. The total flux per pole is 0.05 Wb. The resistance of the armature is 0.3 Ω. When connected to a 240 V supply, at a particular load the armature current is 20 A. Calculate: (a) the speed under these conditions, (b) the value of flux per pole required to increase the speed to 15 rev/s, all other conditions remaining unchanged.

[(a) 10.83 rev/s, (b) 0.036 Wb]

7   Draw simple circuit diagrams illustrating (a) the series connection, (b) the shunt connection and (c) the compound connection.

8   Describe typical applications for DC machines which are: (a) shunt connected, (b) series connected and (c) compound connected.

9   Sketch and justify by using the relevant theory the speed/armature current and torque/armature current characteristics for motors connected as in P.8 (a), (b) and (c) above.

10 A DC shunt motor has an armature circuit resistance of 0.6 Ω and a shunt field resistance of 250 Ω. It is connected to a 250 V supply. On no load the input current is 2.5 A and the speed is 1500 rev/min. When fully loaded the input current is 10 A. Calculate the value of back e.m.f. generated and the speed of the motor on full load.

[no load $E_1 = 249.1$ V, $k\phi = 9.964$, on load $E_2 = 244.6$ V, new speed = 24.55 rev/s (1473 rev/min)]

11 The torque developed by the armature of a DC motor is 95 Nm when the armature current is 50 A. What torque would be developed when the armature current is 25 A assuming (a) the motor is shunt connected and the field remains constant and (b) the motor is series connected and the effects of saturation can be ignored?

[(a) 47.5 Nm, (b) 23.75 Nm]

12 Explain why it is necessary to employ a starter to start a DC motor. Sketch and explain the action of one type of starter. Why is it possible to achieve shorter run-up times especially with a heavy load involved as in the case of a motor used in a traction application by using an electronic starter as opposed to a stepped-resistance starter?

13(a) A DC motor has an armature resistance of 0.3 Ω. It is to be started using a resistance starter. The supply voltage is 250 V. Determine the value of starting resistance for the armature circuit so that the current at standstill does not exceed 25 A.

(b) If, during the run-up, at one speed the back e.m.f. has risen to 125 V, to what value should the starting resistance be reduced to restore the original value of current?

[(a) Add 9.7 Ω, (b) Add 4.7 Ω]

14 What special relationship must exist between armature coils and field poles in order to achieve good commutation? What design features are incorporated in many machines to assist in commutation?

15 What is the difference between 'armature reaction' and 'reactance voltage'? What design feature in a DC machine may be used to overcome the effects of both of them?

16 The effect of armature reaction in a DC machine is to twist the magnetic field around the armature. Explain how this may result in an overall reduction in magnetic flux from the main poles. How will such a reduction affect the performance of (a) a shunt motor and (b) a shunt generator?

17 List and briefly discuss the losses which occur in a DC machine.

18 A DC shunt motor has the following parameters:
Armature resistance = 0.7 ohms, shunt field resistance = 200 ohms.

The supply voltage is 200 V.
The motor is run on no load when the total input current is 4.5 A.
Determine the combined total friction, windage and iron losses of the motor.

[Accountable losses total 208.6 W. Friction etc. by difference = 691 W]

19   A DC motor has a total power input of 25 kW. The field loss is 400 W. The armature losses total 1.3 kW. The efficiency of the motor is 0.892 p.u. Determine the value of the output power and the combined friction, windage and iron loss.

[Output = 22.3 kW, friction etc. = 1 kW]

20   A loaded DC shunt machine has the following parameters:

$R_a = 0.3$ Ω; $R_f = 160$ Ω; Supply voltage = 240 V.
Total current drawn from the supply = 35 A
At this loading the efficiency of the motor = 0.78 p.u.

Determine: (a)   The input power
(b)   The useful shaft output power
(c)   The combined value of all the friction, windage, iron and brush losses.

[(a) 8400 W, (b) 6552 W, (c) 1151 W]

21   A DC series motor runs at a speed of 2650 rev/min when drawing an armature current of 5 A. The motor is loaded such that the armature current increases to 25 A. Given that the combined resistance of the armature and series field = 0.5 Ω and the friction and windage losses of the motor are 600 W, determine: (a) the speed and (b) the efficiency of the motor at this load.

[(a) 516.7 rev/min, (b) 0.916 p.u.]

22   A DC shunt-connected machine is driven at 1500 rev/min. It is on open circuit; it has no externally connected load; and it develops a terminal voltage (V) of 400 V when driven at 1500 rev/min. The armature resistance of the machine is 0.35 ohms and shunt field resistance 50 ohms.

Determine: (a)   The speed of the machine when operating as a motor and drawing a total current from a 400 V supply of 50 A.
(b)   The efficiency of the motor under these conditions if friction, windage and iron losses total 800 W.

[As a generator, $I_f = 8$ A, $E = 402.8$ V, $k\phi = 16.1$. As a motor, $E = 385.3$ V, $n = 23.91$ rev/s, efficiency = 0.769 p.u.]

23   A series-connected DC motor has combined armature and series field resistance of 0.6 ohms. When connected to a 400 V DC supply it draws a current of 28 A and runs at 600 rev/min under specific loading conditions. At this load,

friction and windage total 300 W. (a) Determine for this condition (i) the shaft output torque, (ii) the efficiency of the motor.
(b) The torque is now reduced so that the current input falls to 12 A when the friction and windage losses rise to 1 kW. Determine (i) the new speed of the motor, (ii) the output torque and (iii) the efficiency.

[(a) $k\phi = 38.32$, power at shaft = 10430 W, $T = 166$ Nm.
(b) New $k\phi = 16.42$, (i) $n = 23.92$ rev/s, (ii) 24.7 Nm, (iii) 0.774 p.u.]

Multiple choice questions

1 In a DC shunt motor the total input current is I_L amperes and the field current is I_f amperes.
The relationship between terminal voltage V and back e.m.f. E is:

A $V - (I_L - I_f)R_a = E$
B $V + (I_L - I_f)R_a = E$
C $V - (I_L + I_f) = E$
D $E - (I_L - I_f) = V$.

2 The armature resistance of a DC shunt generator is 0.5 Ω. It delivers 15 A to a load at terminal voltage 250 V. The field current = 1 A. The value of the generated e.m.f. is:

A 257 V
B 242 V
C 258 V
D 243 V.

3 The two ends of a coil in a lap winding are connected to segments on the commutator:

A one pole pitch apart
B which are adjacent
C on opposite sides of the armature.

4 Through a wave wound armature there are:

A as many parallel current paths as there are poles
B four parallel paths
C two parallel paths.

5 A lap-wound armature and a wave-wound armature both have the same number of conductors. Each is fitted in a six-pole DC generator and both are driven at the same speed. The terminal voltage on no load of the lap machine is 200 V. The no-load terminal voltage of the wave-wound machine would be (closely):

A 200 V
B 66 V
C 1200 V
D 600 V.

6 A DC series motor develops a torque of 60 Nm with an armature current of 10 A. Increasing the load on the machine such

that the torque is 120 Nm would cause the armature current to change to approximately:

A   20 A
B   14.2 A
C   40 A.

7   Armature reaction in a DC machine is:

A   the torque produced on the field poles by the rotating armature
B   the induced voltage in the armature coils as the current reverses
C   the effect of the magnetic field set up by the armature currents on the main field
D   the induced voltage in the armature coils created by the interpoles.

8   The effect of armature reaction in a DC motor is to:

A   reduce the no-load speed
B   increase the no-load speed
C   cause the reduction in speed which occurs as the motor is loaded to be less than expected at high loading
D   cause the speed to reduce dramatically as full load is marginally exceeded.

9   Commutating or interpoles are fitted to DC motors in order to:

A   control the speed
B   overcome the effects of reactance voltage and armature reaction
C   allow the machine to carry momentary overloads since being series connected an increase in current will increase the available field
D   allow the motor to be reversed without undue sparking at the commutator.

10   The hysteresis loss in the armature of a DC machine is a function of:

A   the maximum flux density
B   the thickness of the laminations used
C   the resistivity of the steel
D   the speed of rotation.
Select *two* of the factors given.

11   Eddy current losses in the armature of a DC machine are reduced by:

A   reducing the resistance of the steel used
B   using thin strips or laminations
C   using grain oriented steel
D   increasing the working flux density.

4 Induction motors

Summary

This chapter looks at the construction and action of the three-phase single- and double-cage induction motor. It deduces the relationship between slip, frequency, number of poles and speed of rotation. The torque/slip relationship is analysed. Starting methods for induction machines are described. Uses for induction motors are discussed.

The induction motor is an alternating current motor fed from a single or a three-phase supply. We shall only consider the three-phase machine in this chapter, the single-phase machine being essentially a small power device used generally in domestic appliances and where a three-phase supply is not available. It has a drooping speed/torque characteristic very similar to that in a DC shunt motor starting from a no-load speed slightly less than the synchronous value. There are basically two types:

1 *cage machines*, which have short-circuited bars in slots on the rotor and need no external connections to the rotor. This is a relatively cheap method of construction which is very robust.
2 *wound-rotor machines*, which carry windings on the rotor and need slip rings to provide a connection to external resistance banks.

The latter arrangement provides the means to change the torque/speed characteristic whilst the former can only run at essentially constant speed, with no simple means of speed variation.

Synchronous speed

Before examining the action of the induction motor we must briefly digress to look at a synchronous machine which runs at what is called *synchronous speed*. Looking back to Chapter 3, Figures 3.18 and 3.19, we see a DC machine with four poles with an armature carrying a lap winding. Picture the armature winding as in Figure 3.19 with the same conductor numbering 1,1', 2,2' and 3,3' and repeated as 4,4', 5,5' and 6,6' without its commutator but stationary and arranged as in Figure 4.1. The four poles are

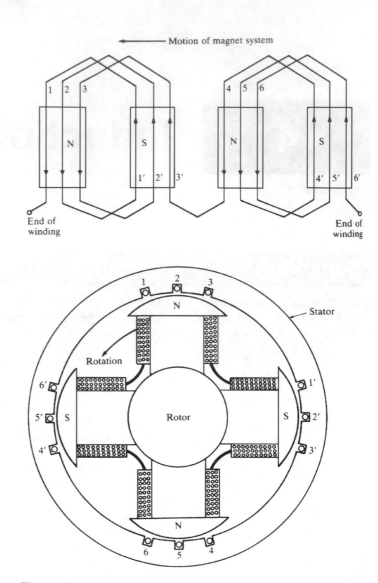

Figure 4.1

shown but now mounted on a rotor with the pole faces reshaped so as to fit inside what is called the *stator*. There is still relative motion between conductors and field but in this arrangement it is the field which moves. The absence of the commutator means that this machine can only be associated with alternating currents.

If we consider any conductor on the stator it will link with the magnetic flux of each of the rotor poles, two north poles and two south poles, during one revolution. The direction of the induced e.m.f. in each conductor will reverse as it links with alternate poles and the voltage appearing at the output terminals will be alternating similar to that developed by the slip ring machine already examined in Chapter 3, Figure 3.6 but with the coil stationary and field moving. There will be two complete cycles of output per revolution of the rotor. Therefore, driving the rotor at 25 revolutions per second will cause the output frequency to be 50 cycles per second (50 Hz).

Test your knowledge 4.1

(a) What is the synchronous speed of a three-phase machine with 12 poles when connected to a 60 Hz system?
(b) How many poles does a particular synchronous machine have if it is found to be running at 12.5 rev/s when connected to a 50 Hz system?

In the synchronous machine there will always be the same number of groups of conductors as there are poles. Looking at Figure 4.1, if the stator were effectively cut in half so removing conductors 4, 5, 6, 4′, 5′ and 6′ together with their associated poles and the remaining windings and poles were expanded into a circular form once more (Chapter 3, Figure 3.19 wrapped round into a cylindrical form), a two-pole machine would result as shown in Figure 4.2(a). During one revolution each conductor would link with one north pole and one south pole giving one cycle of output.

Generally, $\quad n = \dfrac{f}{P}$ rev/s

where $\quad n =$ speed in revolutions per second

$f =$ frequency in cycles per second (Hertz)

$P = pairs$ of magnetic poles.

Adding more groups of conductors and more poles while keeping the arrangement symmetrical makes a range of lower speeds possible. Steam turbines need to run at high speeds so that a one-pole-pair rotor is used. Some low head water turbines run at low speeds and 15–20-pole-pair machines are found. This is also the case with synchronous and induction motors; the more pole pairs, the slower will the machine run. The arrangement of a three-pole-pair machine is shown in Figure 4.2(b).

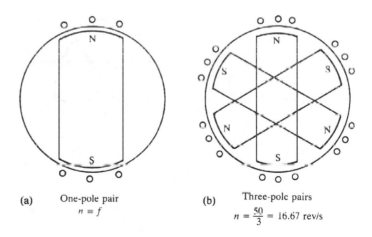

(a)   One-pole pair
$n = f$

(b)   Three-pole pairs
$n = \dfrac{50}{3} = 16.67$ rev/s

Figure 4.2

The rotating magnetic field

A three-phase machine carries three identical windings on its stator so that the magnetic pole system links with each in turn to induce three voltage waveforms displaced by 120°. (See also Chapter 1, Electricity supply.)

When the stator of a machine carrying a three-phase winding is energized from a three-phase supply a rotating magnetic field of constant amplitude is created which rotates within the stator at synchronous speed. We have two relationships: (1) driving a magnetic field system associated with a three-phase winding with an engine gives rise to a three-phase output and (2) feeding a three-phase winding with this supply gives rise to a rotating magnetic field.

The production of this field is demonstrated in Figure 4.3. Three stators for a one-pole-pair machine are shown which have their phase windings reduced to the simplest possible form of one coil per phase per pole. Beneath the stators the waveforms of the supply currents are drawn.

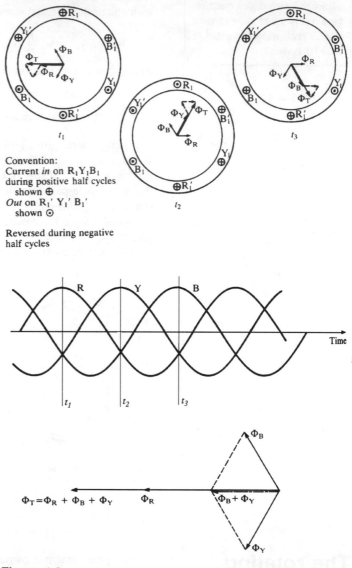

Convention:
Current *in* on $R_1 Y_1 B_1$
during positive half cycles
  shown $\oplus$
Out on $R_1' Y_1' B_1'$
  shown $\odot$

Reversed during negative
half cycles

Figure 4.3

At an instant in time t_1, the red phase current is at a maximum value going positive while the other two phase currents are one half their maximum values going negative. A positive-going current will enter R_1 and leave R_1'. Using the right hand screw or grip rule this current will cause a north pole to be created at right angles to the coil diameter, pointing to the left of the stator as drawn. The negative-going (reverse direction) yellow current will enter Y_1' and leave Y_1. This will create a north pole at right angle to the coil diameter but since this current is only one half of the red current the flux and the phasor representing it are correspondingly shorter. The same procedure is adopted for the blue phase. Adding the flux phasors gives a total flux which has a magnitude of 1.5 times that for the red phase alone. The addition is shown to a larger scale beneath the waveforms.

Two other instants in time t_2 and t_3 are examined in the same way. The flux phasor representing the total flux is constant in magnitude. It is rotating in a clockwise direction in synchronism with the movement along the waveforms. When next the red current has its maximum value the phasor will have arrived back in its original position. This relationship may be demonstrated at any point on the waveform.

Construction of the induction motor

The three-phase induction motor has a stator which is identical to that of the synchronous machine. The one-pole-pair, two-pole-pair and three-pole-pair arrangements of one phase winding on the stator are shown in Figures 4.1 and 4.2. In each case two more identical windings will be added, equally displaced around the circumference. Although the induction motor does not have poles on its rotor, it is described as a one-, two- or P-pole-pair machine according to the arrangement of windings on its stator. Replacing the induction-type rotor with a rotor with the corresponding number of magnetic poles would transform it into a synchronous machine, no other changes being required.

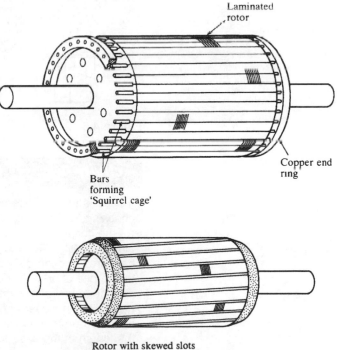

Figure 4.4

Labels: Laminated rotor; Copper end ring; Bars forming 'Squirrel cage'; Rotor with skewed slots cast aluminium bars and end rings

The induction motor cage rotor is built up from laminations of low reluctance steel, slots being provided to carry copper bars. The bars protrude at each end and a large cross-sectional-area copper ring with holes drilled to suit is pressed on to the bars at each end. The bars are brazed into their holes to give a good electrical connection. The arrangement of bars and rings is sometimes referred to as a *squirrel cage* since it resembles the exercise wheel used by small rodents kept in cages. The original motor was described by Nicola Tesla in 1889 and in Russia, so the story is told, squirrels could be seen employing such wheels at the time. A cage machine with the left hand shorting ring partly sectioned is shown in Figure 4.4. In a cheaper form of construction frequently found in smaller motors, the rotor laminations are assembled and placed in a mould. Molten aluminium under considerable pressure is injected into the mould which fills the slots forming bars and end rings simultaneously.

The wound rotor is also laminated and carries a three-phase, star-connected winding. The starts of the phase windings are joined together to form a star point while the three finishes are connected to slip rings which are shrunk on to the shaft but insulated from it. The slip ring brushes provide a means of connecting external resistance banks. The arrangement is shown in Figure 4.5.

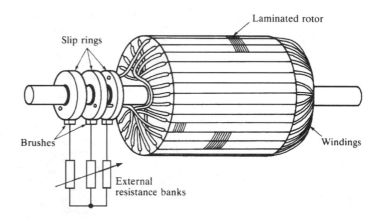

Figure 4.5

Activity 4.1

Inspect a three-phase cage induction motor with a view to checking: (a) the bearings used, (b) the method of cooling, (c) the construction of the rotor and (d) the form of the stator winding. This may be possible using a motor which has already been disassembled for the purpose or by taking one apart as a group. Generally the drive-end cover may be gently prised off after removing four or six nuts around the periphery of the cover which are either tightened on to studs in the stator casing or rods which run the length of the case to nuts at the other end. Pulling off the end cover removes the drive-end bearing at the same time. Use gentle force, possibly tapping with a raw-hide or plastic-faced mallet. Seek guidance from a technician or electrical fitter etc. if in doubt.

Consult makers' data. What classes of insulation are available? What is the range of ratings available? What other methods of cooling are available and where might each be used?

Report on your findings.

Action of the induction motor

When the stator of the induction motor is energized from a three-phase supply a magnetic field is produced within the stator and which links with the rotor. This field rotates at synchronous speed (see Figure 4.3). As the field sweeps past the rotor conductors the bars have e.m.f.s induced in them and consequently currents flow. The action is that of a generator. Hence the name induction motor; it functions because of voltages induced in its rotor.

In Figure 4.6, consider that the stator has just been energized and the rotor has not yet had time to start turning. At this instant in time the stator-produced field has a north pole towards the top of the stator as drawn and is rotating in a clockwise direction. The

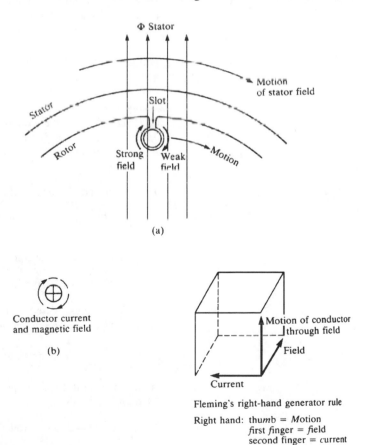

Figure 4.6

rotor bars have a relative motion backwards through the field. Put another way, if the field were stationary the bars would have to be moving anticlockwise to give the same effect. Fleming's right hand generator rule was formulated for DC machines in which the field is stationary while the conductors move. Using the conductor motion through the field it may be deduced that the bar current is away from the viewer. This will set up a clockwise magnetic field around the bar which is shown in isolation in Figure 4.6(b). In its slot this will result in bar field and stator field assisting each other to the left of the bar and opposing each other to the right of the bar. The resulting strong field to the left of the bar will produce a force causing it to move in the same direction as the rotating field.

There are many bars on the rotor all producing torque and the rotor will accelerate. If the rotor ever attained synchronous speed, that is to say if the bars ever moved at the same speed as the rotating field, the induced e.m.f.s and bar currents would fall to zero and there would be no torque. The rotor would slow down, the induced e.m.f.s would reappear and torque would again be produced. It follows therefore that for any particular load applied to the motor there is a speed at which the rotor currents will produce just enough torque to drive that load. With the rotor speed too small the induced e.m.f.s and resulting currents are large enough to produce an accelerating torque. If the rotor speed is too high, the rate of flux cutting, induced e.m.f.s and currents fall so reducing the torque and the rotor slows down. The speed/torque characteristic is therefore a slightly drooping one, an increase in torque causing a small decrease in speed.

In order to generate the rotating field, currents flow in the stator giving rise to copper losses (I^2R_{STATOR}) and iron losses. Rotor currents are matched with balancing currents in the stator. The various currents and losses may be compared directly with those in the transformer.

Slip

The synchronous speed of the rotating field is given by:

$$n = \frac{f}{P} \text{ rev/s}$$

The actual speed of the rotor $= n_R$ rev/s

$$n - n_R = n_S \qquad \text{Where } n_S \text{ is called the slip speed.}$$

The slip speed expressed as a proportion of the synchronous speed is the fractional, percentage or per unit slip S.

$$\frac{n_S}{n} = S \qquad \text{Transposing: } n_S = S \times n$$

A machine with a synchronous speed of 25 rev/s and running at 24 rev/s has a slip speed:

$$n_S = (25 - 24) = 1 \text{ rev/s} \qquad S = 1/25, 4\% \text{ or } 0.04 \text{ p.u.}$$

Now since $n - n_R = n_S$

$$\text{Transposing: } n_R = n - n_S$$
$$= n - Sn$$
$$= n(1 - S)$$

Frequency of the rotor currents

Considering the instant just as the supply has been connected to the stator and before the rotor has commenced to turn, the stator and rotor are in effect the primary and short-circuited secondary of a transformer. At this instant:

$$n_R = 0, \text{ and since } n - n_R = n_S, n_S = n \text{ and } S = 100\%$$

Since, in a transformer, there can be no frequency change between primary and secondary, the frequency of the currents in the rotor is the same as the frequency of the supply to the stator.

$$f_R = f \text{ Hz}$$

Where f_R = frequency of the rotor currents

f = frequency of the supply

Consider now the rotor to be running at synchronous speed. The rotor bars are running at exactly the same speed as the rotating stator field. There can be no induced e.m.f. as already discussed. The only way such a condition can be sustained without external driving force is for the rotor to be energized with direct current. It would become a synchronous motor. The frequency of the rotor currents would be zero. For this condition:

$$n_R = n \text{ and again, since } n - n_R = n_S, n_S = 0 \text{ and } S = 1$$

Between the limits $S = 0$ and $S = 1$, the frequency of the rotor currents falls from the supply value to zero.

$$f_R = S \times f$$

Example 4.1

An eight-pole induction motor connected to a 50 Hz supply runs with a slip of 5%. Determine: (a) the speed of the motor and (b) the frequency of the currents in the rotor.

(a) $\qquad n = \dfrac{50}{4} = 12.5$ rev/s (8 poles = 4 pole pairs)

$$n_S = Sn$$

$$= 5\% \text{ of } 12.5$$

$$= 0.625 \text{ rev/s}$$

$$n_S = n - n_R$$

$$\therefore 0.625 = 12.5 - n_R$$

$$n_R = \textbf{11.875 rev/s}$$

Or directly:
$$n_R = n(1 - S) = 12.5(1 - 0.05) = 11.875 \text{ rev/s}$$

(b) $\qquad f_R = Sf$

$$= 5\% \text{ of } 50$$

$$= \textbf{2.5 Hz}$$

Power and torque

In Figure 4.7, work done per revolution $= 2\pi r F$ Nm.
Work done per second $= 2\pi r F n$ joules per second $= 2\pi n T$ watts (see also Chapter 3).

Suppose the torque supplied by the rotor of an induction motor $= T$ Nm. (This value includes the externally driven mechanical load and friction and windage on the rotor itself due to its bearings and cooling fan etc.)

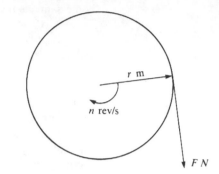

Figure 4.7

The rotor is turning at n_R rev/s

$\therefore$ power developed by the rotor $= 2\pi n_R T$ watts

$$= 2\pi(n - n_S)T$$

$$= 2\pi n(1 - S)T$$

$$= (1 - S)2\pi n T \ \text{W}$$

Now: the torque developed by the rotor = torque supplied by the stator. This must be true since if the stator torque were less than the demand of the rotor, it would slow down until the currents in stator and rotor were sufficient to result in equilibrium at a new, lower speed. If the stator torque were greater than that demanded of the rotor it would speed up until once again equilibrium would be restored.

The stator magnetic field revolves at synchronous speed n rev/s

$\therefore$ stator power $= 2\pi n T$ W($=$ power transferred to the rotor in the magnetic field)

Hence, power supplied to the rotor : power developed by the rotor

$$2\pi n T : (1 - S)2\pi n T$$

which is a ratio $1 : (1 - S)$

The difference between the two quantities:

$$2\pi n T - (1 - S)2\pi n T = S \times \text{input power to the rotor}$$

The quantity S, the slip, is associated with the rotor loss both in speed and in power. It follows that the lower the value of slip, the lower the rotor losses and the higher the efficiency of the motor.

Example 4.2

A ten-pole induction motor connected to a 50 Hz supply operates with a slip of 4.5 per cent when the shaft power = 25 kW. Determine:

(a) The torque developed by the motor.
(b) The power being transferred from the stator to the rotor.

(a) $n = 50/5 = 10$ rev/s   4.5% = 0.045 p.u.

Power developed by the rotor $= 2\pi n_R T$ watts

$$= (1 - S)2\pi n T$$

$$\therefore 25\,000 = (1 - 0.045)2\pi \times 10 \times T$$

$$T = \frac{25\,000}{(1 - 0.045)2\pi \times 10}$$

$$= \textbf{416.64 Nm}$$

(b) Power supplied by stator $= 2\pi n T$

$$= 2\pi \times 10 \times 416.64$$

$$= 26\,178 \text{ W}$$

Or since power developed by the rotor $- (1 - S)$ × input to rotor

$$\text{Power input} = \frac{\text{Power developed by rotor}}{(1 - S)}$$

which yields the same result without having first to calculate the torque value.

Test your knowledge 4.4

The input to the stator of a 50 Hz, three-phase, six-pole induction motor is 10 kW. For a rotor speed of 16 rev/s, determine:

(a) the rotor loss and
(b) the torque developed by the rotor.

The torque/slip characteristic

Consider the wound-rotor induction motor. This makes the concepts easier to understand but the principles apply to all induction motors.

With the rotor at *standstill*, i.e. $S = 1$, the e.m.f. induced in the rotor winding per phase, i.e. between the star point and one slip ring, $= E_R$ volts.

The value of this induced voltage is directly proportional to the rate at which the stator flux is cut, that is to say, directly proportional to the slip S. As S approaches zero, the rotor approaches synchronous speed and the e.m.f. falls towards zero.

Rotor voltage at any value of slip $S = S \times E_R$ volts/phase.

Let the resistance of the rotor winding be R_R Ω/phase.

The rotor winding also has a large inductance since it comprises a number of turns on an iron core. Let the inductance be L_R H/phase.

At *standstill* the reactance $X_R = 2\pi f L_R$ Ω.

At slip S, the frequency of the rotor currents falls to Sf Hz when:

$$\text{Reactance} = 2\pi(Sf)L_R = S \times \text{standstill value} = SX_R \text{ Ω}$$

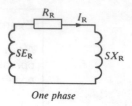

One phase

Figure 4.8

The electrical equivalent circuit for one phase of the rotor is as shown in Figure 4.8.

$$\text{Rotor current } I_R = \frac{V}{Z} = \frac{SE_R}{\sqrt{[R_R^2 + (SX_R)^2]}}$$

$$\text{Power loss in the rotor} = I^2 R_R = \frac{S^2 E_R^2 R_R}{[R_R^2 + (SX_R)^2]} \text{ watts}$$

But loss $= S \times$ power input to the rotor.
Transpose: power loss$/S =$ power input to rotor.
   Divide the loss expression by S.

$$\text{Power input to the rotor} = \frac{SE_R^2 R_R}{[R_R^2 + (SX_R)^2]} \text{watts per phase}$$

Using this expression, the torque/slip characteristic of a typical induction motor may be deduced.

Example 4.3

A three-phase induction motor has a synchronous speed of 25 rev/s, standstill rotor voltage of 100 V/phase, rotor resistance of 0.25 Ω/phase and standstill rotor reactance of 1.5 Ω/phase. Use these parameters to plot the torque/slip characteristic for the machine.
   At standstill $S = 1$

$$I_R = \frac{1 \times 100}{\sqrt{0.25^2 + (1 \times 1.5)^2}} = \frac{100}{1.521} = 65.76 \text{ A}$$

$$\begin{aligned} \text{Rotor loss} = I^2 R_R &= 65.76^2 \times 0.25 \\ &= 1081.08 \text{ W/phase} \\ &= 3 \times 1081.08 \text{ W in three phases} \\ &= 3243 \text{ W total rotor loss} \end{aligned}$$

The rotor loss shows up as heat and cooling arrangements must be provided sufficient to keep the rotor cool.

$$\text{Input power} = \frac{\text{Power loss in the rotor}}{S}$$

$$= \frac{3243}{1} = 3243 \text{ W}$$

$$\text{Input power} = 2\pi n T$$

$$3243 = 2\pi \times 25 \times T$$

$$T = \frac{3243}{2\pi \times 25} = 20.65 \text{ Nm}$$

At $S = 0.5$

$$I_R = \frac{0.5 \times 100}{\sqrt{(0.25^2 + (0.5 \times 1.5)^2)}} = 63.24 \text{ A}$$

Rotor losses $= 3(63.24^2 \times 0.25) = 3000$ W
Input $=$ loss$/S = 3000/0.5 = 6000$ W

$$T = \frac{6000}{2\pi \times 25} = 38.19 \text{ Nm}$$

The reader might care to work through the calculations for $S = 0.25$ ($T = 58.8$ Nm) and for $S = 0.05$ ($T = 35$ Nm). At $S = 0.1667$

$$I_R = \frac{0.1667 \times 100}{\sqrt{(0.25^2 + (0.1667 \times 1.5)^2)}} = 47.14 \text{ A}$$

$$\text{Power input} = \frac{3 \times 47.14^2 \times 0.25}{0.1667} = 10\,000 \text{ W}$$

$$\text{Torque} = \frac{10\,000}{2\pi \times 25} = \mathbf{63.67}$$

The results of these calculations are plotted in Figure 4.9. The first observation from the graph is that the starting torque of the motor (when $S = 1$) is very low being only 20.65 Nm. The motor will be capable of running up to

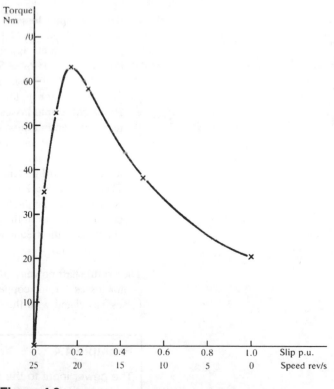

Figure 4.9

A six-pole, three-phase, 50 Hz cage induction motor has the following parameters:

$$E_R = 65 \text{ V/phase}$$

$$R_R = 0.15 \text{ }\Omega\text{/phase}$$

$$X_R = 0.75 \text{ }\Omega\text{/phase}$$

From a consideration of the rotor equations determine:

(a)   the speed for maximum torque
(b)   the value of the maximum torque
(c)   the standstill (starting) torque.

Losses in induction motors

speed only if the applied torque at standstill is less than this value.

Because of the low starting torque induction motors are most often started with no applied load, this being applied when the motor is up to full speed, for example when driving machine tools where cutting operations only commence when the work is at full speed and centrifugal pumps which are started with their discharge valves closed, these being opened to commence pumping when at full speed.

The second observation is the connection between R_R and SX_R at the maximum torque condition. Maximum torque occurs at $S = 0.1667$ p.u.

Investigating the relationship between R_R and X_R

$$\frac{R_R}{X_R} = \frac{0.25}{1.5} = 0.1667$$

Maximum torque occurs at a value of slip given by R_R/X_R. This is clearly not a rigorous proof of the relationship, but the result is a valid one.

- **Stator losses.**   The stator of the induction motor is made up of laminations of steel in the same way as the core of a transformer. It suffers the same hysteresis and eddy current losses as the transformer. Currents flowing in the stator windings give rise to heat losses as $I^2 R_{\text{STATOR}}$ All these losses must be provided by the power source.
- **Rotor copper losses.**   There are copper losses in the rotor bars or winding which have already been discussed. These are $I^2 R_R$ which is equal to slip multiplied by the rotor power input. $I^2 R_R = S \times$ Rotor input; both quantities being per phase. This means that the rotor develops a power $= (1 - S) \times$ Rotor input.
- **Rotor rotational losses.**   The rotor is driven through the air in the casing and in most cases the shaft carries a fan to blow air across the rotor and stator to keep it cool. The shaft provides the fan power and necessary torque to overcome the frictional loss. There will be friction in the bearings and shaft end seals. These losses are known as 'friction and windage'. The rotor is made from laminated steel and suffers hysteresis and eddy current losses as in the stator. However, these are usually quite small since the frequency of the alternations is slip frequency. $f_R = S \times f$.

The useful shaft power is therefore the input power to the stator − (stator losses + rotor copper losses + friction, windage and iron losses associated with the rotor).

Example 4.4

The power input to the stator of a six-pole, 50 Hz cage induction motor is 54 kW when it is operating at a slip of

4 per cent. The stator losses are 1450 W and rotor friction, windage and iron losses are 650 W.
Determine: (a) the motor speed, (b) the copper loss in the rotor, (c) the power output at the shaft, (d) the output torque and (e) the overall efficiency of the motor.

(a) $n_R = (1 - S)n = (1 - 0.04) \times 50/3 =$ **16 rev/s**

(b)
$$\text{Input to stator} = 54 \text{ kW}$$
$$\text{Loss in stator} = 1.45 \text{ kW}$$
$$\therefore \text{ Input to rotor} = 54 - 1.45 = 52.55 \text{ kW}$$
$$\text{Copper loss in the rotor} = S \times \text{power input}$$
$$= 0.04 \times 52.55$$
$$= \textbf{2.102 kW}$$

(c) Power output = power input to rotor − copper loss
− friction, windage and iron losses

$$= 52.55 - 2.102 - 0.65$$

$$= \textbf{49.8 kW}$$

(d) Using output quantities, the shaft power leaves the machine at the actual speed of the machine, i.e. 16 rev/s

$$2\pi \times 16 \times T = 49\,800$$

$$\text{Torque developed} = \frac{49\,800}{2\pi \times 16} = \textbf{495 Nm}$$

(e) Efficiency $= \dfrac{\text{shaft output}}{\text{stator input}} = \dfrac{49.8}{54} = $ **0.92 p u** **(92%)**

Starting methods for cage machines

Direct on line

Cage motors may be started by connecting them directly to the supply employing a contactor or circuit breaker. The current input to the motor during the period taken for it to attain full speed is large being typically between five and eight times the normal full load value. Such a large current flowing in the power supply transformer and distribution network can cause an unacceptable reduction in supply voltage to other consumers. In an extreme case it might be envisaged that the supply voltage could fall to such a value as to affect the current input to the motor itself, substantially increasing the run-up time. Despite the extremely large current, the power input is not proportionally large resulting in a poor starting torque. This is due to the poor starting power factor.

$$\cos\phi = \frac{R_R}{Z_R} = \frac{R_R}{\sqrt{R_R^2 + (SX_R)^2}}$$

Using the data from Example 4.3, $R_R = 0.25 \ \Omega$ and $X_R = 1.5 \ \Omega$. Then using $S = 1$ for starting (standstill) conditions:

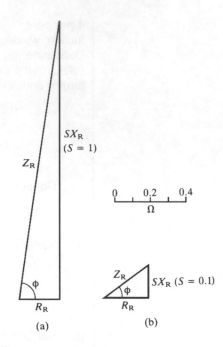

Figure 4.10

$$\text{Power factor at standstill} = \frac{0.25}{\sqrt{(0.25^2 + 1.5^2)}} = 0.164$$

As the motor runs up to speed the value of slip decreases so that (SX_R) decreases causing the power factor to improve. Figure 4.10 shows the impedance triangles for the induction motor for two conditions.

In some circumstances direct-on-line starting is acceptable, for example a motor in a factory which is fed from its own, probably large, supply transformer. The high current is supplied without causing undue voltage drop or affecting other consumers. Elsewhere, where the distribution system is less robust, small motors with correspondingly low starting current requirements and a short run-up time may be started direct on line; the limit on size can only be determined by the effect on the electrical system.

Star-delta starting

The motor is started with the stator windings connected in star then when the motor is close to full speed, they are reconnected in delta.

In Figure 4.11 the main contactor is closed together with contactor S which provides a star point. The motor runs up to speed connected in star and then contactor S is opened. Contactor D is closed which joins winding starts and finishes in sequence and each winding across line voltage so creating the delta connection. Contactors S and D have to be interlocked to ensure that it is impossible to close both simultaneously. This method of starting reduces the current inrush and starting torque in proportion.

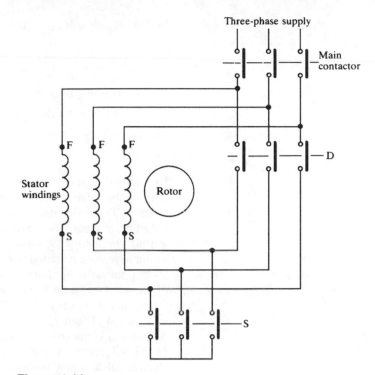

Figure 4.11

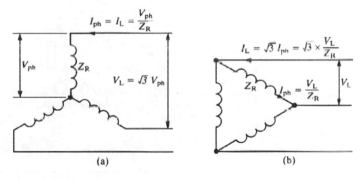

$$I_{ph} = I_L = \frac{V_{ph}}{Z_R}$$

V_{ph} Z_R

$$V_L = \sqrt{3}\, V_{ph}$$

$$I_L = \sqrt{3}\, I_{ph} = \sqrt{3} \times \frac{V_L}{Z_R}$$

Z_R $I_{ph} = \frac{V_L}{Z_R}$ V_L

(a)   (b)

Figure 4.12

With reference to Figure 4.12. Given a motor with standstill impedance Z_R Ω/phase

$$\text{When started in star the phase current} = \frac{V_{phase}}{Z_R}\ \text{A}$$

In a star system, $I_{phase} = I_{line}$. Starting the motor in delta means that the line voltage is applied directly to each phase when the phase current would be:

$$I_{phase} = \frac{V_{line}}{Z_R}$$

In a delta-connected system line and phase currents are connected by $\sqrt{3}$ as were the voltages in the star-connected system (see also Chapter 1).

In delta $I_{\text{line}} = \sqrt{3} I_{\text{phase}}$ and $V_{\text{line}} = \sqrt{3} \times V_{\text{phase}}$

$$\therefore I_{\text{line}} = \sqrt{3} \times \frac{V_{\text{line}}}{Z_R} = \sqrt{3} \times \frac{\sqrt{3} \times V_{\text{phase}}}{Z_R} \text{ A}$$

$$\frac{I_{\text{line}}(\text{starting in star})}{I_{\text{line}}(\text{starting in delta})} = \frac{V_{\text{phase}}/Z_R}{\sqrt{3} \times \sqrt{3} \times V_{\text{phase}}/Z_R}$$

$$= \frac{1}{3}$$

By starting the motor in star the current inrush is reduced to one third of the value which would be drawn by the motor if started in delta. This reduction in current brings about a corresponding reduction in input power and hence torque. Using this method of starting, the starting torque of the motor is only one third of what it would have been if started connected in delta. The delta-connected starting torque is itself small (see Figure 4.9) so that this method of starting must be limited to motors which can be started off load.

The star-delta method of starting using contactors as shown in Figure 4.11 can give rise to current and voltage surges as contactor S is opened and contactor D is closed. This is because of the inductance of stator and rotor in which, as the currents change, voltages may be induced giving rise to excessive currents sometimes larger than those which would have occurred using

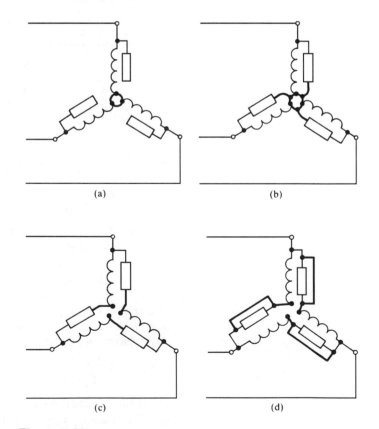

(a)

(b)

(c)

(d)

Figure 4.13

direct-on-line starting. A better arrangement for large motors is shown schematically in Figure 4.13.

Figure 4.13(a) shows the supply connected to the motor while a contactor forms a star point at the centre of the diagram. Figure 4.13(b) shows resistors connected in parallel with the motor windings which are still in star. Such connections are made by using additional contactors. Figure 4.13(c) shows further contactor operation resulting in each of the motor windings being connected in series with a resistor giving a delta formation. Finally in Figure 4.13(d) the resistors are shorted out leaving the windings alone in delta. The method gives smooth transition from the star to the delta configuration without disconnecting the windings at any time.

Auto-transformer starting

With star-delta starting the standstill current inrush was reduced by reconnecting the windings so that the voltage on each was less than the normal running value when connected in delta. The reduction in voltage could only be that value dictated by the phase/line voltage relationship, i.e. $\sqrt{3}$ or 1.732.

With the auto-transformer method the motor remains connected in delta at all times while the reduction in voltage is achieved by the use of a transformer.

In Figure 4.14, the auto-transformer is star-connected. Its output voltage per phase is V_s so that the output line voltage which is applied to the motor is $\sqrt{3}V_s$ volts which is called V_m. When starting, the contactor is switched to the position shown in Figure 4.14. When the motor approaches full speed the contacts are moved to the top (run) position. The motor is now connected directly to the supply, the transformer being redundant. The transformer tapping may be adjusted to give any starting voltage

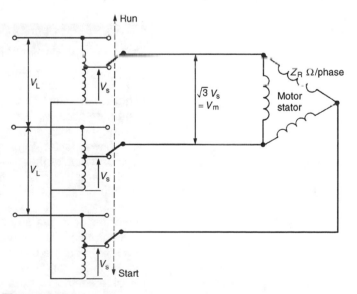

Figure 4.14

required. As with star-delta starting it should be observed that a reduction in voltage with its corresponding reduction in current gives a reduced starting torque.

$$\text{In the motor, } I_{\text{phase}} = \frac{V_m}{Z_R} \text{ A and } I_{\text{line}} = \frac{\sqrt{3} \times V_m}{Z_R} \text{ A}$$

$$\text{The transformer voltage ratio} = \frac{V_L}{V_m} : 1$$

$$\therefore \text{current ratio (inverse to voltage ratio)} = \frac{V_m}{V_L} : 1$$

$$\text{Hence, line input current to the transformer} = \frac{V_m}{V_L} \times \frac{\sqrt{3}V_m}{Z_R}$$

$$= \frac{\sqrt{3}V_m^2}{V_L Z_R} \text{ A}$$

$$\text{When started direct on line, the input current} = \frac{\sqrt{3}V_L}{Z_R} \text{ A}$$

(See Figure 4.12.)

$$\frac{\text{Line input to the auto-transformer}}{\text{Line input to motor, direct on line}} = \frac{\dfrac{\sqrt{3}V_m^2}{V_L Z_R}}{\dfrac{\sqrt{3}V_L}{Z_R}}$$

$$= \frac{V_m^2}{V_L^2}$$

With this type of starting, the current inrush to the auto-transformer is V_m^2/V_L^2 times that which would flow in the same motor if it were started direct on line. Once again, a reduction in input power results in a reduction in torque so that this method of starting can only be used with motors starting off load.

Test your knowledge 4.6

When started direct on line a three-phase induction motor has an initial current inrush of 650 A when the starting torque is 150 Nm. What will be the values of starting current and torque if the motor is started using an auto-transformer with step-down ratio 2.5:1?

Activity 4.2

Carry out tests on a squirrel-cage induction motor and report on your findings. The motor should be started using either a star-delta starter or an auto-transformer starter. Instrumentation will be such as to measure the line input current, line voltage, total power input and optionally the input power factor. Either a three-phase wattmeter can be used or a single-phase type with the current element in the red line and the voltage element fed from the red line down to neutral when the total power will be three times the indicated value allowing for any meter constants. If a power factor meter is not available, the power factor may be determined as a power/volt ampere ratio ($W / \sqrt{3}V_L I_L$). Load the machine using a friction or electrical dynamometer and measure the speed using either a portable contact or fixed tachometer.

Start the motor on no load using the starter. Observe the current swings on the ammeter but the changes will be rapid and no meaningful values can be read. Is there an interlock to prevent the starter going to the direct-on-line position with the motor at standstill? What type of interlock is it, mechanical or electronic?

Increase the load on the machine in steps using the dynamometer until full-load current is reached, recording the readings on all instruments at each stage.

Plot graphs of: (a) torque/speed and slip (this will correspond to the first section of the characteristic in Figure 4.9), (b) efficiency/load, $2\pi n_R T$ watts (horizontal) and (c) power factor/load.

Comment on the shapes of the three curves. When considering the power factor curve, why is the power factor poor on low loads when from a consideration of the rotor parameters, at low values of slip the rotor appears to be very nearly pure resistance? Why is it only possible to determine the first part of the characteristic in Figure 4.9? What other test could be carried out to fix one further point on the curve and what might prevent you from doing so in practice? Why, when buying an induction motor, would it be best to buy one which matches your power requirement closely rather than one, say, twice as large giving considerable scope for carrying an overload?

Look at a copy of the relevant British Standard which covers the motor you have used. What does it say about permitted overload? What types of enclosures are mentioned? Investigate IP numbers for electrical equipment. (These are concerned with the standards of enclosure necessary to keep out water etc.) Obtain maker's data and incorporate any relevant details in your report.

Starting the wound rotor induction motor

See Figure 4.5 and the section on 'Construction'.

Using this type of motor provides for starting against a large load torque while limiting the current inrush. It also considerably improves the starting power factor as compared with that achieved by the methods for starting cage machines described above.

In Example 4.3 the relationship between rotor resistance R_R, rotor reactance SX_R and torque was briefly examined. Maximum torque is developed by an induction motor when the relationship $R_R = SX_R$ is satisfied. The brushes running on the slip rings provide the means whereby sufficient resistance may be added to the rotor circuit to balance the equation at any value of slip. This will cause maximum torque to be developed at any particular speed between standstill and its normal slightly sub-synchronous value.

In Figure 4.15 four possible torque/slip curves are shown. The first envisages the rotor circuit resistance being increased to match the value of the standstill reactance.

Then, since $R_R = SX_R$ for maximum torque and $R_R = X_R$, maximum torque must occur at $S = 1$, i.e. at standstill. The motor

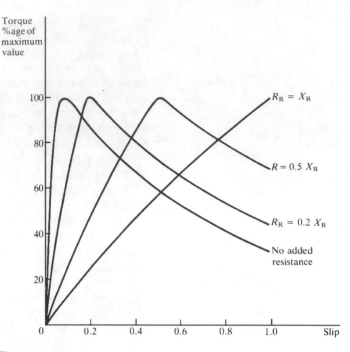

Figure 4.15

develops its maximum torque at starting and the characteristic is seen to be falling towards zero as the slip decreases, that is to say as the motor speed increases.

With added external resistance to bring the total up to a value equal to one half of the rotor standstill reactance, for R_R to equal SX_R, the slip S must be 0.5. The same general shape of characteristic as shown in Figure 4.9 may be plotted for this case with its maximum torque at $S = 0.5$.

For R_R set to a value 0.2 times the standstill reactance, the maximum torque occurs at $S = 0.2$ and this curve is also shown in Figure 4.15. The final curve with no added resistance is that already shown in Figure 4.9.

Using this method to start the motor, the stator is energized with as much resistance in the rotor circuit as is required either to give the required starting torque or to limit the inrush current to a specific level. As the motor speed increases the rotor resistance is eliminated in stages using contactors or a manual arrangement similar to a face-plate starter for a DC machine. A measure of speed control may also be achieved by leaving sufficient resistance in the rotor circuit to produce a given torque at a substantially sub-synchronous speed. This is not to be recommended, however, because of the amount of energy dissipated in the resistance bank with subsequent low efficiency.

Example 4.5

A 50 Hz, four-pole, wound-rotor induction motor has the following parameters:

A four-pole, 50 Hz, wound-rotor induction motor has the following parameters:

$$E_R = 55 \text{ V/phase}$$

$$R_R = 0.05 \text{ } \Omega/\text{phase}$$

$$X_R = 0.45 \text{ } \Omega/\text{phase}$$

Resistance is added to the rotor circuit in order to produce maximum torque at standstill. Determine:

(a) The value of resistance required per phase.
(b) The value of the starting torque.
When the motor is running up and has reached a slip of 0.5 determine:
(c) The value of torque at this slip.
(d) The efficiency of the rotor. (Power developed by the rotor/power input to the rotor.)
(e) For the motor running at $S = 0.5$ determine the overall efficiency of the motor if for this condition the stator loss $= 700$ W and the friction, windage and iron loss for the rotor $= 250$ W.

Rotor resistance $R_R = 0.1$ Ω/phase
Rotor standstill reactance $= X_R = 0.5$ Ω/phase
Rotor standstill voltage $E_R = 65$ V/phase.

(a) For starting purposes the rotor circuit resistance is increased to match the standstill rotor reactance. For the instant of starting ($S = 1$) calculate: (i) the magnitude of the rotor current per phase, (ii) the rotor power factor and (iii) the torque developed by the rotor.

(b) During the run-up the rotor circuit resistance is reduced to 0.25 Ω/phase. Recalculate the values in (i), (ii) and (iii) above for $S = 0.5$.

(a) (i) $$I_{rotor/phase} = \frac{SE_R}{\sqrt{[R_R^2 + (SX_R)^2]}} = \frac{65}{\sqrt{[0.5^2 + 0.5^2]}}$$

$$= \textbf{91.92 A}$$

(ii) $$\cos\phi = \frac{R_R}{Z_R} = \frac{0.5}{\sqrt{(0.5^2 + 0.5^2)}} = \textbf{0.707}$$

(See Figure 4.10.)

(iii) Rotor loss $= 3I^2R_R^2$

$$= 3 \times 91.92^2 \times 0.5 = 12\,674 \text{ W}$$

At $S = 1$, loss $=$ input. $n = 50/2 = 25$ rev/s

Torque developed $= \dfrac{12\,674}{2\pi \times 25} = \textbf{80.68 Nm}$

(b) (i) $$I_{rotor/phase} = \frac{0.5 \times 65}{\sqrt{[0.25^2 + (0.5 \times 0.5)^2]}} = \textbf{91.92 A}$$

(ii) $$\cos\phi = \frac{0.25}{\sqrt{[0.25^2 + (0.5 \times 0.5)^2]}} = \textbf{0.707}$$

(iii) Rotor loss $= 3 \times 91.92^2 \times 0.25 = 6337$ W

Rotor input $= \dfrac{\text{loss}}{S} = \dfrac{6337}{0.5} = 12\,674$ W

Torque developed $= \dfrac{12\,674}{2\pi \times 25} = \textbf{80.68 Nm}$

This example demonstrates that by making reductions in the rotor circuit resistance as the motor runs up to speed, the current, power factor and torque may be maintained at substantially constant values.

Double-cage machines

In order to combine the advantages of the cage machine in that it needs no electrical connections to its rotor with those of the wound-rotor machine which develops high torque at starting, various different arrangements of rotor bars have been developed. One of these arrangements involves the use of two separate cages and is called the *Boucherot* rotor.

From Faraday, we have the formula:

$$\text{induced e.m.f.} = \frac{N\Delta\phi}{\Delta t} \text{ volts}$$

We have also a formula for the induced e.m.f. in terms of inductance:

$$e = L\frac{\Delta i}{\Delta t} \text{ V}$$

$$\therefore L\frac{\Delta i}{\Delta t} = \frac{N\Delta\phi}{\Delta t}$$

and cross multiplying:

$$L = \frac{N\Delta\phi \times \Delta t}{\Delta i \times \Delta t} = \frac{N\Delta\phi}{\Delta i} \text{ H}$$

where N = number of turns

$\Delta\phi$ = flux change in webers

Δt = time interval in seconds

Δi = change in current

L = inductance measured in henry

Thus, if a current changes by an increment Δi and the resulting change in flux linkages $N\Delta\phi$ is known, the inductance of the circuit can be found.

A conductor which is surrounded with a high-permeability material such as steel has a high inductance since, for a given conductor current, a large magnetic flux is created or, expressed in another way, for a given current change, the field change is large.

A conductor situated totally or partly in air will have a much lower inductance since air has a much lower permeability.

Figure 4.16(a) shows the arrangement of slots in a double-cage rotor. The bottom cage is situated deep down in the steel and the flux paths are largely in the steel, having only to cross the narrow slot. Its inductance is therefore high. This cage is constructed using materials with low resistance. The flux paths associated with the current in the upper cage have to pass through a large length of air, leaving the rotor, crossing the air gap into the stator slots. This cage has low inductance and is formed using materials with relatively high resistance. The general arrangement of the bars and shorting rings is shown in Figure 4.16(b).

Using typical values of resistance and reactance, the action of the rotor may be investigated.

For the top cage, $R_R = 1 \, \Omega$, $X_R = 1 \, \Omega$.
For the bottom cage, $R_R = 0.2 \, \Omega$, $X_R = 5 \, \Omega$.

Consider the rotor at standstill, $S = 1$

The impedance of the top cage $= \sqrt{(1^2 + 1^2)} = 1.414 \, \Omega$

The impedance of the bottom cage $= \sqrt{(0.2^2 + 5^2)} = 5.004 \, \Omega$

Because of its lower impedance much more current will flow in the top cage than in the bottom cage. It will be observed that since

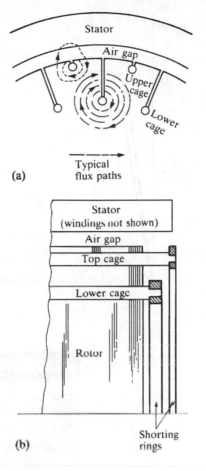

(a)

(b)

Figure 4.16

$R_R = X_R$ for the top cage it will produce its maximum torque at standstill.

Now consider the speed to have risen such that $S = 0.1$.

$$\text{The impedance of the top cage} = \sqrt{[1^2 + (0.1 \times 1)^2]}$$

$$= 1.005 \ \Omega$$

$$\text{The impedance of the bottom cage} = \sqrt{[0.2^2 + (0.1 \times 5)^2]}$$

$$= 0.54 \ \Omega$$

At this value of slip twice as much current flows in the bottom cage as in the top cage. As the speed increases the current in the high-resistance cage decreases while that in the low-resistance cage increases. The characteristic of the motor is the combination of that of a wound-rotor machine for starting changing over to that of a normal, low-resistance cage motor for running. The total characteristic is developed in Figure 4.17.

Motors are available with three cages employing various methods of interconnecting them resulting in differing characteristics to suit particular loads.

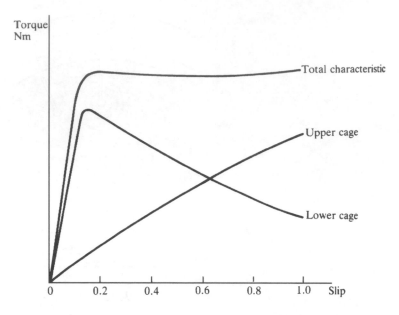

Figure 4.17

Activity 4.3

Conduct tests on a wound rotor induction motor. Instrumentation will be required to measure line voltage and current to the machine, power input and the speed of the shaft. A friction dynamometer will be required since the test entails locking the shaft to determine the standstill torque.

If a resistance-type starter is provided, measure the values of the various stages: all in; with one stage switched out etc. down to the final position when zero resistance will be in circuit. If no starter is provided it will be necessary to use a suitably rated rheostat; the initial value of this, probably less than 10 Ω, will need to be determined by experiment.

Lock the rotor by taking up the friction dynamometer adjustment to a high level. Put all the resistance into the rotor circuit and energize the stator. Record the torque reading on the dynamometer and the stator input current. Lower the rotor resistance by one stage on the starter or by a small increment on the rheostat. Observe what has happened to the torque. If it has increased then initially there was too much resistance in the rotor circuit for maximum torque. If it has decreased then the resistance is now too small. The resistance may be decreased in steps to obtain the values of standstill torque until the stator current is at its full load value. Do not keep the motor stationary for too long a period since at standstill it will rapidly overheat because its fan is not running.

Having established the value of resistance to produce maximum torque, energize the stator and slowly release the dynamometer allowing the motor to run up. Record torque, speed, line current and input power at, say, four intervals. Plot the torque/speed characteristic for this value of added resistance. Reduce the resistance by an increment and repeat the test, better now starting at full speed and braking down to standstill if this can be achieved without overloading the stator. Plot characteristics as shown in Figure 4.15 as far as possible. For the high values of resistance it will be possible to plot the whole characteristic but for lower values only the first part starting from a slip close to zero (no load) up to the point where maximum torque is approached will be possible.

From the results extract data relating to a particular torque, say 0.7 of the maximum, and determine the efficiency and power factor at which the motor operates for each resistance stage used at this torque. Comment on the relationship between efficiency and added resistance for this particular torque.

Write a report detailing your findings and stating why you would recommend, or not, this method for starting an induction motor.

Uses for induction motors

Cage induction motors are used wherever a drive with virtually constant speed is required. This includes fans, pumps, compressors, conveyor drives and the range of machine tools which employ a gearbox to provide the work drive as in, for example, the centre lathe or milling machine.

Wound-rotor machines are used to start high-torque loads and to provide limited speed control. An example is the passenger lift. The rotor resistance is cut out in stages giving controlled acceleration of the cage which could be heavily loaded. Permanent low-speed running of a load could be achieved using this method but is very inefficient and other methods possibly employing solid-state control would be preferable.

Multiple-cage induction motors are available which have two separate stator windings or the possibility of regrouping the stator windings into series/parallel groups. These motors have two distinct permanent running speeds. A use is found in driving boiler forced draught fans. On loading up to, say, one half of full load the low-speed winding is used, fine control being by damper. For loads beyond this and up to full load, the high-speed winding is used, again fine control being by damper.

A number of methods of controlling induction motors have been developed employing solid-state electronics, in particular the thyristor and the gate-turn-off thyristor (GTO). However, every development which makes this technology cheaper does the same for the control of direct current motors, which can be readily controlled from a crawl up to full speed so that the future choice of variable-speed machine types is not at all clear.

Problems on induction machines

1 Describe the construction of three types of induction motor rotor.

2 What is the main advantage of the induction motor over the equally rated DC motor?

3 Why is it impossible for the induction motor to run at synchronous speed without external assistance?

4 When the stator of an induction motor is supplied from a three-phase system a magnetic field is produced which has two important properties. Explain how this magnetic field is produced and state what the two properties are.

5 Why is it important that induction motors carrying steady loads operate at low values of slip?

6 Describe two methods employed in induction motors to generate high torque at standstill.

7 What is the speed of an induction motor with eight poles when connected to a 50 Hz supply and running with a slip of 4 per cent.

[6 rev/s]

8 What is the value of slip for a six-pole induction motor connected to a 50 Hz supply when it is running at 16 rev/s?

[4%]

9 What is the frequency of the currents in the rotor of an induction motor with four poles when connected to a 60 Hz supply and running at 28.5 rev/s?

[3 Hz]

10 A six-pole induction motor connected to a 50 Hz system runs with 3 per cent slip. It develops a torque of 53 Nm. Determine: (a) the power being transferred across the air gap from the stator to the rotor and (b) the shaft power.

[(a) 5550 W, (b) 5384 W]

11 Sketch the torque/slip characteristics of induction motors with the following parameters:
(a) $R_R = 0.1$ Ω, $X_R = 0.4$ Ω
(b) $R_R = 0.5$ Ω, $X_R = 0.5$ Ω
(c) $R_R = 0.25$ Ω, $X_R = 0.6$ Ω.

12 Explain why the rotor power factor of a cage machine improves as the rotor runs up from standstill to near synchronous.

13 What are the advantages of using a wound-rotor induction motor over a cage machine?

14 Describe two methods of starting cage induction motors which reduce the current inrush. What are the possible disadvantages of these methods?

15 A six-pole, 50 Hz, three-phase induction motor has a shaft output power of 6.3 kW when running with 4.5 per cent slip. The friction, windage and iron losses in the rotor are 260 W and the stator losses are 850 W.

Determine: (a) the input power to the rotor
    (b) the copper loss in the rotor
    (c) the power input to the stator
    (d) the output torque at the shaft and the torque developed by the rotor before taking account of the losses from the shaft
    (e) the efficiency of the motor under these conditions.

[(a) Rotor develops 6.56 kW, input = 6.87 kW, (b) 0.309 kW, (c) 7.72 kW, (d) o.p. torque corresponds to 6.3 kW at 15.92 rev/s = 63 Nm, Torque developed = 65.6 Nm, (e) 0.816 p.u.]

16 An eight-pole induction motor connected to a 50 Hz supply has the following parameters:

Standstill rotor voltage per phase, $E_R = 53$ V
Rotor resistance per phase, $R_R = 0.08$ Ω
Rotor standstill reactance per phase, $X_R = 0.42$ Ω.

Determine for a slip of 5 per cent, (a) the rotor power factor, (b) the torque developed by the shaft. If the stator loss = 600 W and the rotor friction windage and iron loss combined = 250 W at this load, determine (c) the output torque and (d) the efficiency of the motor.

[(a) 0.967, (b) 62.7 Nm, (c) 59.38 Nm, (d) 0.8 p.u.]

17 An induction motor develops a starting torque of 146 Nm when started direct on line with its stator windings connected in delta. What would be the expected values of starting torque if started (a) using the star/delta method and (b) using an auto-transformer giving a line voltage ratio of 415/375 V at starting?

[(a) 48.7 Nm, (b) 119.2 Nm]

18 What advantages has a multiple-cage induction motor over a single cage machine. Sketch the overall characteristic of a multiple-cage machine and state a typical application.

19 What link is there between the development of variable-speed induction motors and DC motors? How does this link make the choice of drive difficult at times?

20 What is the only feature of an induction motor which would need to be changed in order to have a machine running at synchronous speed?

Multiple choice questions

1 The synchronous speed of a three-phase machine depends upon:

A the supply voltage
B the connection of the stator, star or delta

C the arrangement of the stator winding in series/parallel groups

D the provision of a magnetized rotor.

2 The shaft speed of an induction motor is equal to:

A synchronous speed + slip speed

B synchronous speed multiplied by (1 + fractional slip)

C synchronous speed multiplied by (1 − fractional slip)

D frequency of the supply multiplied by the number of pole pairs.

3 The main benefit of an induction machine is that:

A it has a wide range of running speeds

B it needs no electrical connection to its rotor as in the case of the DC machine

C it is especially efficient at low speeds

D it runs at synchronous speed irrespective of loading.

4 A three-phase supply is used in conjunction with the three-phase induction motor (as opposed to using a single-phase set up) because:

A the three-phase arrangement produces a torque three times as great as the single phase

B the magnetic field produced is of constant size and rotates at constant angular velocity

C it produces a very high starting magnetic field.

5 The speed of a 60 Hz, six-pole, three-phase induction motor running with 5 per cent slip is:

A 10 rev/s

B 20 rev/s

C 19 rev/s

D 9.5 rev/s.

6 A 50 Hz, four-pole, three-phase induction motor runs at 24 rev/s. Its per unit slip is:

A 0.52

B 0.04

C 0.05.

7 The frequency of the currents induced in an eight-pole, 50 Hz, three-phase induction motor is 2 Hz. The speed of the rotor is:

A 6.25 rev/s

B 12.5 rev/s

C 12 rev/s

D 6 rev/s.

8 The main advantage of the wound-rotor induction motor over the cage machine is:

A the electrical connections are made to the slip rings instead of to the stator

B the torque can be adjusted to suit start-up conditions

C although it has poor starting power factor it is much better than the cage machine when at full speed

D it produces very high efficiency during starting.

9 The starting torque of a cage machine is:

A high

B better than a double-cage machine

C low

D dependent upon the addition of resistance to the rotor circuit.

10 The rotor of a cage machine rotates:

A in the opposite direction to the synchronous field in accordance with Lenz's law

B in the same direction as the synchronous field and at a higher speed so that the bars continue to cut this field

C in the same direction as the synchronous field but at a slower speed so as to cut the field

D either direction according to how the starter is operated.

11 Cage machines are very suitable for use where:

A virtually constant speed is required

B variable torque is required

C variable speed is required

D constant torque independent of speed is required.

12 In a double-cage machine:

A the upper cage has a lower resistance than the lower cage

B the upper cage has a higher resistance than the lower cage

C both cages have the same resistance but different reactances.

13 With a cage machine maximum torque occurs at:

A high values of slip

B low values of slip

because the ratio of R_R/X_R in the motor rotor is:

C large

D small.

Choose A or B together with C or D to make a correct statement.

14 When starting a cage machine in star the torque compared with starting in delta is:

A $1/\sqrt{3}$ as large

B $1/3$ as large

C $\sqrt{3}$ times greater

D 3 times greater.

15 When starting a cage motor using an auto-transformer with a step-down ratio 1/2, the torque on starting compared with direct on line is:

A half the magnitude
B two times greater
C one quarter the magnitude
D four times greater.

5 Materials and their applications in the electrical industry

Summary

This chapter examines materials used for carrying current and the extent to which grouping, ambient temperature and current-carrying capacity influence conductor type and size. The necessity for heat sinks is explained. Losses in the ferrous cores of transformers and chokes are analysed and soft and hard steels compared. Reasons for failure of dielectric materials is examined.

Conductor materials for overhead lines and underground cables

The best electrical conductor known is silver but this is far too expensive and rare to provide all the conductor material required by the electrical industry. Next in order of conductivity come copper and aluminium and these are the most important current-carrying materials used in cable and line manufacture. Other materials used are cadmium copper, phosphor bronze and for some high-voltage low power links, galvanized steel.

The conductivity of both copper and aluminium falls very rapidly with very small additions of alloying elements so that they are generally used pure. In the case of aluminium the mechanical strength is improved by using a stranded conductor with steel strands at the centre.

Figure 5.1 shows a stranded conductor and the make-up will be 1 strand of steel plus 6 strands of aluminium; 7 strands of steel plus 12 strands of aluminium, or by adding a further layer of aluminium strands, 7 of steel plus 30 of aluminium.

The conductivity of the whole is taken as that of the aluminium alone since steel has a very high impedance and the current flows almost exclusively in the aluminium.

1+6+12

Figure 5.1

The total strength of the cable is normally 50 per cent greater than that of the same conductivity copper cable. The result of using steel cores is to produce cables which are smaller for a given tensile strength than copper or, although larger for a given resistance, much stronger than copper. Table 5.1 shows some of the electrical and mechanical properties of copper and aluminium.

Table 5.1

Property	Copper	Aluminium
Weight	87 200 N/m³	26 700/m³ (Aluminium 0.306 times copper)
Resistivity	1.73×10^{-8} Ωm (0.975 as good as silver)	2.87×10^{-8} Ωm (Aluminium 1.64 times copper) (0.585 as good as silver)
Strength	Ultimate, 320 MN/m²	Ultimate 144 MN/m² (Aluminium 0.45 times copper)
Flexibility	When annealed, quite good. Used hard for cables and overhead lines when it must be stranded to give the required flexibility	Very flexible, can be used solid in cables
Jointing	Soldered ferrules for cables. Compression joint on overhead lines	Compression type on overhead lines. Crimped lugs for cable terminations. Can be soldered or welded using special fluxes but generally more difficult than copper
Resistance to corrosion	Excellent. Virtually none in most cases over very long periods	Poor when in contact with other metals and in particular copper and copper-bearing alloys Special sleeves and fittings required. On overhead lines there is limited deterioration; much of the original 1933 grid is still operational in its original form

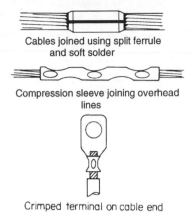

Cables joined using split ferrule
and soft solder

Compression sleeve joining overhead
lines

Crimped terminal on cable end

Figure 5.2

Performance of insulating materials

Loss angle δ

In order to prevent current leaking away from cables and other conductors, they need to be suspended on or surrounded by insulating materials. For example, overhead power lines are insulated from earth by air along their length and by porcelain or glass insulators from the steel supporting towers, whilst inside, leakage from power cables is prevented by enclosing them entirely in plastic materials or paper. Such insulating materials are known as *dielectrics*.

A dielectric is a material which has a very high resistivity in comparison with a conductor such as copper and aluminium. However, no dielectric is perfect and there will always be some current leakage. This leakage current, flowing in its resistance, is in phase with the voltage and so represents a power loss. In addition, consider how a capacitor is formed. Any conducting surface close to another but separated from it by an insulator has capacitance. A capacitor for use in electronic circuits often comprises two very thin aluminium foils separated by waxed paper or mica dielectric. Now since the current-carrying conductor of an overhead line or within a power cable is near to earth and probably near to the return conductor of the circuit, capacitance is created from conductor to conductor and from each conductor to earth, the insulating material around the conductor being the dielectric.

The current which flows in the circuit because of its capacitance has a value given by the circuit voltage divided by the capacitive reactance of the system.

$$I_C = \frac{V}{X_C} = \frac{V}{1/\omega C} = V\omega C \text{ amperes}$$

The current flowing in the circuit because of the imperfect dielectric $= I_R$ amperes.

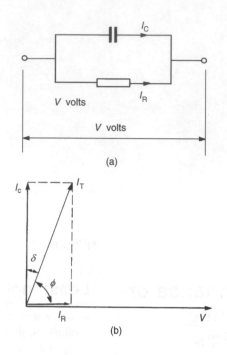

Figure 5.3

The working equivalent circuit and the quantities are shown in Figure 5.3(a).

The phasor diagram may be constructed as in Figure 5.3(b). I_C leads the circuit voltage by 90° whilst I_R is in phase with the voltage. The total circuit current I_T leads the voltage by an angle ϕ. Since I_R is usually very small, ϕ approaches 90°. The angle δ is the difference between 90° and ϕ, and is known as the loss angle since it is only present because of the dielectric loss. In a perfect capacitor the current would lead the voltage by exactly 90°.

Good cable dielectrics have very small values of loss angle, being only a few minutes of arc. Since insulation resistance decreases rapidly as temperature increases, it follows that the loss angle δ increases with temperature (I_R increases). As δ increases, the loss increases, giving rise to more heat and further temperature rises. It can be envisaged that this could lead to failure if the temperature were allowed to rise too far. This is particularly important in very high voltage cables where the heat produced by this effect becomes very nearly as great as that due to conduction losses in the conductor (I^2R losses) See later in this chapter, 'Heating of cables'.

Breakdown voltage

If a sufficiently high voltage is applied to any insulating material it will break down and become conducting. Particular materials have uses as dielectrics for a particular range of voltages. The breakdown strength of a dielectric depends on its thickness but is not proportional to it. This means that if a sample of insulating

---

Test your knowledge 5.1

A sample of insulating material sandwiched between two metal plates has a loss angle of 0.25° and a capacitance of 0.3 μF (see Figure 5.3). What is the power loss in the material when a voltage of 10 kV at 50 Hz is applied across the two plates?

material is subjected to increasing voltage it will eventually punc-
ture and possibly set on fire. Let us suppose that the sample tested
was 1 mm in thickness and failure occurred at 5 kV, using twice
the thickness would not result in a breakdown voltage of 2×5 kV
but more likely only 7 or 8 kV. Increasing the thickness further
gives even less increase, 3 mm thickness breaking down at perhaps
only 10 kV.

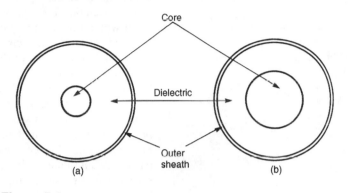

Figure 5.4

In cable design it is found that the diameter of the conductor
has a great influence on breakdown voltage. The smaller this is,
the more likely the cable is to fail. In Figure 5.4, cable (a) will fail
at a substantially lower voltage than cable (b). See also Chapter 1,
Figure 1.5 and observe that the diameter of the conductor in the
extra-high-voltage cable has been made large by the inclusion of an
oil duct. A similar cross-sectional area of copper could be provided
by omitting the duct and reducing the diameter. This increase in
diameter and the provision of oil to improve the insulating properties of the paper dielectric makes it possible for the cable to
support very high voltages.

Insulating materials for cables

For underground cables operating at up to 3.3 kV, the most
common insulating material is polyvinyl chloride (PVC). Street
mains have either copper or aluminium conductors. These are
PVC-insulated, then three or four such conductors are laid up to
form a cable which is sheathed overall in PVC, and surrounded
with steel wire armouring if required (see also Chapter 1,
Figure 1.5).

House wiring usually comprises stranded or solid copper
conductors which are insulated with PVC (see Figure 5.5).

(a)   (b)

Stranded copper Two stranded copper
conductor   conductors PVC
               insulated

PVC insulation   Central single strand
               earth wire

Figure 5.5

Polychloroprene (PCP) is used to insulate copper conductors in farming installations where resistance to ammonia and other such corrosives is required.

The insulating material in mineral-insulated cables is magnesium carbonate which spaces solid copper conductors within a copper tube. It has a number of advantages amongst which is the ability to operate at red heat – a property which no other cable insulating material can match.

The insulation for high-voltage power cables is predominantly paper with oil or gas filling under pressure. At extra high voltage it is essential that the dielectric be absolutely uniform with no small air bubbles or particles of dirt present. It would be at these small imperfections, since they would be weaker than the perfect dielectric, that breakdown would occur. A small arc would form, burning the insulation producing carbon and other conducting combustion products. The process would then be continuous, the arc getting progressively larger until the cable failed explosively. In paper-insulated cables the conductors are wrapped in many layers of paper tape and then oil impregnated under vacuum to completely soak the paper and fill the minute gaps present with oil. Using this construction, all air bubbles and dirt are excluded. When operating, many oil-filled cables are kept under pressure so that any tendency for the oil to migrate out as the cable heats up is prevented. Within recent years plastic insulation called cross-linked polyethylene (XLPE) has been developed for use at voltages up to 275 kV. The construction is very similar to the paper type, poly-thene replacing paper, but there is no oil and therefore no provision of oil ducts within the cable. It promises to be cheaper to install and maintain than the pressurized paper type but as yet there is little of it in service.

Activity 5.1

Conduct high-voltage breakdown tests on a range of available materials.

Carry out the following tests in a high-voltage test cage.

1   Breakdown of air. Using conical or small diameter spherical electrodes investigate the breakdown voltages for gaps varying from 1 mm up to the maximum at which breakdown can be achieved. To measure the breakdown voltage on a specific gap, increase the applied voltage very gently keeping a watch on the voltmeter. When gap breakdown occurs the gap voltage will fall to an extent dependent on the equipment. Record the value of voltage observed immediately before the fall. If available, a recording voltmeter can be used when the highest voltage reached will be apparent. Plot a curve of breakdown voltage (vertical) against the gap in millimetres. Comment on its shape.

2   Repeat the tests using electrodes with greater diameter. How does the diameter of the test electrode influence the results?

3 Using flat-faced electrodes of as large a diameter as practicable test the breakdown values of any available insulating materials – PVC, laminated circuit board etc. – quoting the results in kV/mm and stating the actual thickness tested, e.g. the breakdown of a 1.5 mm thickness at 6 kV gives a breakdown or dielectric strength $= 6/1.5 = 4$ kV/mm on a 1.5 mm specimen. Note that with a 3 mm specimen the result might well be 3.5 kV/mm. Testing a double thickness of the material will give a fallacious result since it is influenced by air pockets and dirt, finger marks etc. on the touching surfaces. If the samples tested are of differing thickness no absolute comparison between them can be made. Record the mode of failure; is there a hole punched in the material, does it set on fire etc.?

Use any reference source available to you to research dielectric failure; why do thicker specimens in general fail at a lower stress than thinner ones? Look for any British Standards concerned with materials and high-voltage testing. Investigate the standard test used on transformer oil. If you can avail yourself of a standard test cell, this test can be carried out.

Cable construction

At low voltages the single-stranded conductor with PVC insulation is most commonly used. These are drawn into steel or plastic conduit (Figure 5.5(a)).

In domestic and commercial situations the cable shown in Figure 5.5(b) comprising two stranded or solid copper conductors with a central earth wire is often used. These cables are either run on the surface, inside capped trunking or buried in the walls with suitable safeguards against drilling through them at a later date. In hazardous situations, mostly in industry, the mineral insulated cable (MICS) is used and this is also run on the surface or buried (Figure 5.6).

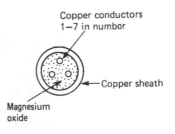

Figure 5.6

For connecting appliances to the supply at low voltages via a plug and socket, the three-core flexible cable is employed (two cores only if the appliance can run earth free, as in the case of double-insulated equipment). This comprises fine stranded copper wire which can withstand flexing, PVC insulation and belt,

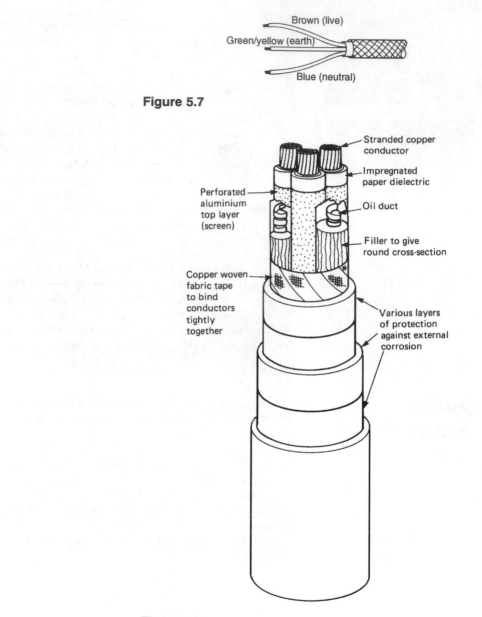

Figure 5.7

Figure 5.8

covered overall with a braided protective fibre or plastic sheath (Figure 5.7).

For frequencies higher than that of the mains, the flat twin or coaxial construction is adopted. (See Conductor screening and coaxial cable, below.)

The construction of extra-high-voltage cables is illustrated in Figure 5.8 and Chapter 1, Figure 1.5.

Multi-core cables are used where many circuits are required to run over the same route between a pair of terminal points, for example for the control of equipment, metering and indication and transmission of condition information such as temperature, pressure, position and power consumption. Power to the equipment concerned is usually fed through a separate cable to minimize the risk of damage to low voltage equipment should a fault occur.

The cables comprise the required number of cores with fibre fillers if necessary to make up a circular cross-section, with a belt of insulation and armouring if required. Each core is numbered or otherwise identified along its length.

The heavy current types in common use are:

1   Stranded copper conductors with paper or fabric insulation with overall paper wrapping, lead covering, steel wire armoured with an anti-corrosion sheath. The cable is terminated at each end in a compound-filled box.
2   Stranded copper conductors with PVC insulation, PVC belt, armoured with PVC anti-corrosion sheath overall.

Each conductor is terminated in a soldered or crimped lug which is bolted to a terminal block from which connections to equipment are made (Figure 5.9).

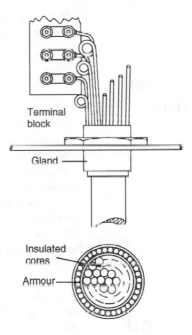

Figure 5.9

Lighter types comprise the required number of single strands of fine copper wire insulated with PVC and covered with a belt of PVC. End connections may be crimped as before or bound and soldered to terminal posts (Figure 5.10).

Multi-core cables for telecommunications have pairs of fine insulated wires twisted together along their length to form circuits, two pairs then being twisted together. Many such groups of four conductors make up a complete cable.

For the interconnection of printed circuit boards, flexible flat cables may be used. The spacing between the conductors is the

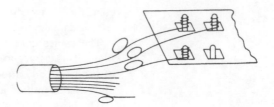

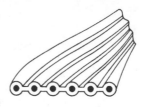

Figure 5.10

Figure 5.11

same as that for the contacts in the edge connectors. Cables with 50 or even more conductors are available (Figure 5.11).

Materials used in electronic circuits

Because of its good conductivity and mechanical strength copper is used extensively to connect electronic components either in the form of insulated stranded wires or strip bonded to an insulated board in the form of a printed circuit. The insulation used is of polythene, polystyrene, polyurethane enamel or air where conductors can be suitably spaced.

Permanent connections between components and interconnecting copper are usually soldered. Where components or boards of components must be capable of removal from an equipment, possibly for repair, some form of sliding contact system is required. The plug and socket for connection of domestic electrical equipment is an example of such a system.

Conductor screening and coaxial cable

A conductor carrying a current sets up a magnetic field around it, the strength of which is proportional to that current. The magnetic field may affect other circuits in its vicinity and to prevent this circuit screening is necessary.

Where a component is to be screened from a steady magnetic field it is surrounded with a material with low magnetic reluctance. In Figure 5.12 the component to be screened is marked C. Without the screen the magnetic field will affect the component. With the screen in position the magnetic lines of force take the low reluctance path which is through the nickel-iron, so leaving the component in a position of zero flux.

Where alternating current is involved the conductor carrying the current, and hence producing the alternating magnetic flux, is surrounded by the screen. The arrangement is in effect a transformer, the conductor being the primary and the screen a short-circuited secondary.

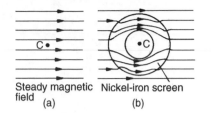

Steady magnetic   Nickel-iron screen
field
(a)   (b)

Figure 5.12

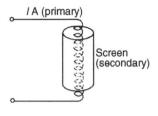

Figure 5.13

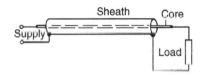

Figure 5.14

The alternating flux induces a voltage in the screen and a current flows which sets up an opposing magnetic flux. The two fluxes are very nearly equal so that the magnetic field outside the screen is virtually zero. (Figure 5.13).

The magnetic field around a cable feeding a circuit is eliminated by using the coaxial construction. This is particularly important at very high frequencies when non-screened cables can cause considerable interference with other circuits.

Current is supplied to its load along the core of the cable and is returned to the supply along the sheath which completely surrounds the core. Since the go and return currents are in opposite directions the magnetic fields produced are in opposite directions and completely cancel each other outside the sheath. (Figure 5.14).

The cable may be formed by using a solid core with a copper tube for the sheath with solid insulation (MICS) or by using a solid or stranded core with a braided outer sheath for flexibility when the insulation is generally polythene.

For many applications air is the best insulation. However, air alone is not always practicable, since the conductors have to be kept to their correct spacing. At frequencies up to 200 MHz open wires may be used employing small spacing elements of non-conducting material as shown in Figure 5.15. An alternative construction is shown in Figure 5.16 in which two parallel conductors are lightly insulated and maintained in their positions using a thin plastic web. This construction is used for audio frequencies, feeding loud speakers from an amplifier, for example,

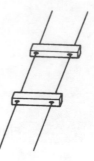

Figure 5.15

Figure 5.16

and at radio frequencies, often being used to connect a VHF (FM) aerial to a receiver. The spacing between the conductors is critical, this having a pronounced effect on the performance of the equipment.

At frequencies up to 3000 MHz the coaxial construction may be used. A compromise between using solid insulation and air is achieved by using either plastic foam containing a substantial amount of air in the form of bubbles or by using a polythene spiral wound on the inner core, when again the insulation is substantially air. This is shown in Figure 5.17.

Figure 5.17

Printed circuits

The electrical connections between circuit components must have low resistance and this can be achieved by using insulated copper wire or bare copper or aluminium strip, the working components then being individually soldered to the conductors. In equipments with many components there are several problems to overcome. One is that of the dry joint, when the soldering looks sound but in fact has not made good contact and then the joint has a high resistance. Another problem with individual hand wiring is that it is difficult to reproduce a circuit accurately in successive equipments. There may be errors in connection and slightly different lengths and routing for wiring taken. This affects not only the resistance but the capacitance of the circuit which has a considerable effect on circuits operating at high frequencies. Each piece of equipment therefore needs individual testing and possible adjustment after manufacture to make sure that it is within specification. Fault finding can be very difficult and it may be necessary to unsolder connections to do tests.

These problems have been largely overcome by mounting the components on an insulating board to a fixed pattern. This has been made possible by the development of transistors and integrated circuits which have low power loss making it possible to place components close together without risk of overheating. The short lengths of conductor between components results in low connection resistance and low capacitance between sections of the circuit. The fixed pattern ensures repeatable values in each and every equipment.

The board must provide mechanical support for the components and the means for their interconnection. The layout of the components is much clearer than with the individually wired equipment and the values of the components are often marked on the board. Now that the precise position of every component is known, the components are often placed in their positions by automatic computer-controlled machines making assembly fast and accurate. Servicing of the equipment is simpler since faulty boards may be replaced to get the equipment working and the components on the faulty boards replaced at a central workshop equipped with diagnostic equipment.

Modern rigid boards are made of phenolic or epoxy resins reinforced with woven glass fabric or paper. They are available up to about one metre square. Flexible boards are available and these are made from polyester film or for high temperature work from PTFE or polypropylene possibly reinforced with glass fibre. Boards for very high frequency operation may be made from Teflon.

Printed circuit boards employ copper foil conductors securely cemented to one of the above laminates. Components may be surface mounted (Figure 5.18) which shows a general arrangement of a four-terminal device with detail of one connection or have tails pushed through drilled holes (Figure 5.19). Surface connections are made by welding or compression whilst through tails are mass soldered by passing the board over, and in contact with the meniscus of, a bath of molten solder.

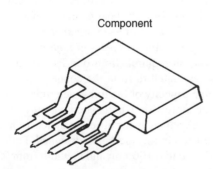
Component

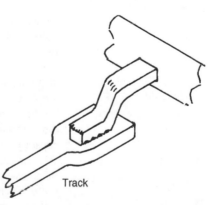
Track

Figure 5.18

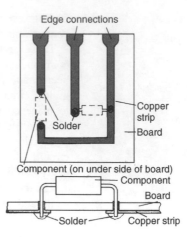

Figure 5.19

To construct a printed circuit board it is necessary to draw a master circuit diagram. Larger areas of foil have to be allowed for parts of the circuit which will carry the heaviest currents. For prototypes this may be drawn directly on the board but for production runs it is drawn considerably larger than the final board to make the work easier and more precise and then photo-reduced to its final size.

Two processes are available for the production of the final circuit.

1 *The subtractive method.* A board is obtained which has foil completely covering one side. The diagram is drawn or printed on the foil using acid-resisting paint and the board is dipped into acid which dissolves the copper which is not protected. Only the required circuit is left on the board and the necessary preparation can be carried out for circuit assembly.
2 *The additive method.* Insulating board is used with no foil covering. The circuit is drawn or printed on the board using conducting paint. Copper is deposited from a plating bath over the treated areas. Other additive methods involve the use of metallic powder and heat or foil strips and mechanical force.

Double-sided boards are commonly produced with circuitry on both sides of the board. Multi-layer boards may be made by using two or more double-sided boards pressed together with a thin layer of insulation between them. Holes or *feed-through* paths are drilled at points where contact is required from one face to another and the surface of the holes plated to form the connection.

The current-carrying capacity of the foil is determined largely by the permissible temperature rise. Charts showing current-carrying capacity and cross-sectional areas for different increases in temperature are consulted when circuits are being designed. Foils vary in thickness from 0.035 mm to 0.106 mm and in width from 0.25 mm up to several millimetres.

As an example a strip 0.07 mm thick and 0.76 mm wide on a single-sided board can carry 3.5 A allowing a 40 °C rise in

Test your knowledge 5.2

A printed circuit board has a strip conductor on its face which is 0.04 mm thick and 2.5 mm wide. The allowable current density is 55 A/mm^2.
(a) What is the permissible current for this conductor?
(b) If the strip were doubled in width would it be permissible to double the current? Give your reason.

temperature but this falls to 2 A when allowing a 10 °C rise, both from 15°.

The cross-sectional area of this strip $= 0.07 \times 0.76$
$$= 0.053 \text{ mm}^2.$$
Allowing a 40°C, rise the permissible current density $= 3.5/0.053$
$$= 65.8 \text{ A/mm}^2.$$
Allowing a 10°C rise, the permissible current density $= 2/0.053$
$$= 37.6 \text{ A/mm}^2.$$

Increasing the strip width gives a less than proportional increase in current-carrying capacity so that current for other strips cannot be deduced by proportion.

The current densities in these very small conductors are considerably greater than those which can be achieved with mains cables since the insulation is very much thinner and the surface area available for cooling is greater per given volume.

To connect the board into the main circuit either plug and socket or edge connectors are used.

Figure 5.20 shows the form of an edge connector. Where large boards are used there may be over 100 connections to be made so that considerable force may be necessary to insert a multi-pin plug into its socket or the board into its edge connector. When using such force it is difficult to determine whether a correct match of the contacts has been made or whether in fact some of them are being badly deformed or crumpled up.

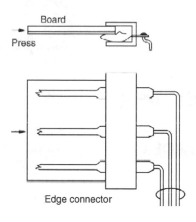

Figure 5.20

There are several patent ways of achieving low insertion pressure followed up by clamping of the board in position. The situation is eased considerably by the choice of contact materials. For example, the pressure needed between gold or platinum contacts to obtain very low contact resistance is only 0.03 times that necessary to obtain the same resistance between brass contacts. The value for nickel and silver is 0.3 times that for brass.

Gold is generally used for low voltage connections of this type since in addition to the low contact pressure required, it does not oxidize or corrode. Even a very thin oxide film can cause a virtual open circuit unless the voltage employed is large enough to break this down. The gold, generally less than 50×10^{-6} m thick, is plated over silver or nickel, the latter combination being especially good in hostile environments.

Interconnection between boards is achieved using insulated wires which are grouped into multi-core cables or merely bunched and neatly carried round the chassis of the equipment. Great care has to be taken to ensure that the capacitance between cores in such close proximity does not affect the circuit performance, that all the connections are sound and that any one circuit does not affect any other as already discussed in the section 'Conductor screening and coaxial cable'.

Activity 5.2

Investigate the construction of a circuit board for a piece of electronic equipment and the method of interconnecting boards.

Inspect a printed circuit board used in a computer, television set or perhaps a radio. What types of components have been used? Are there any indications on the board as values or a code which would assist in the assembly either by hand or automatic machine? Are the components surface-mounted or is the board drilled with soldering on the reverse side? If at all possible visit an assembly plant and check out methods of mass soldering. Look at and comment on the method of board interconnection.

Obtain any available makers' data. Look at suppliers' catalogues to ascertain the range of boards available.

Circuit damage due to overcurrents

Consider a single conductor carrying a current of I_1 amperes and the resulting magnetic field as shown in Figure 5.21.

The magnetizing force at point P is given by:

$$H = \frac{I_1}{\text{length of force line}} = \frac{I_1}{2\pi r} \text{ A}$$

Since the magnetic flux density $B = \mu_0 H$ tesla (strictly in vacuum, but considered true for air and solid insulators)

$$B = 4\pi \times 10^{-7} \times \frac{I_1}{2\pi r} = 2 \times 10^{-7} \times \frac{I_1}{r} \text{ tesla}$$

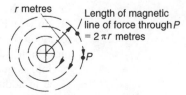

Conductor carrying I_1 amperes

Figure 5.21

A second conductor carrying I_2 amperes running parallel to the first and through point P will suffer a force given by:

$$F = BI_2 \times \text{length of the conductor } l_2.$$

The arrangement is shown in Figure 5.22.

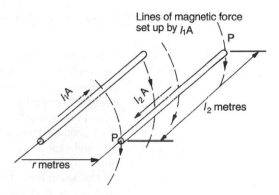

Figure 5.22

Making the length $l_2 = 1$ m, $F = 2 \times 10^{-7} \times \dfrac{I_1 I_2}{r}$ newtons per metre length.

Action and reaction being equal and opposite this must also be the value of the force on the first conductor. With currents in opposite directions as shown there is a repulsive force between the conductors. In single-phase AC and direct-current circuits $I_1 = I_2$ and under normal conditions the forces involved are small. However, when a short circuit fault occurs the currents and forces can be extremely large, sufficient in fact to disrupt the cable or conductor system.

Consider two conductors 4 mm apart carrying (a) a normal current of 20 A and (b) a fault current of 2500 A.

(a) $F = 2 \times 10^{-7} \times \dfrac{20 \times 20}{4 \times 10^{-3}} = 0.02$ newtons per metre run

$$(l_2 = 1 \text{ m})$$

(b) $F = 2 \times 10^{-7} \times \dfrac{2500 \times 2500}{4 \times 10^{-3}} = 312.5$ N/m run

In three-phase circuits each conductor is situated in a magnetic field produced by the currents in the other two conductors and the same effect is produced.

In addition to the forces involved, the heating effect of large currents must be considered.

Power loss $= I^2 R$ watts.

Increasing the circuit current from 20 A to 2500 A will increase the power loss by a factor $(2500/20)^2$ or 15 625 times. It is easy to envisage the effect which this will have on the circuit insulation. Within a few cycles of an AC supply, plastic insulation will melt.

Heating of cables

Since all conductors in normal service have resistance, the passage of current along the conductor gives rise to a voltage drop. The voltage drop is in phase with the current producing it so that the power loss is equal to the product of voltage drop and current flowing.

Consider one core: Volt drop $= IR$ where $R =$ resistance of one core

Power loss = volt drop × current $= IR \times I = I^2R$ watts per core

This power heats the core producing a rise in temperature of the conductor and hence of its insulation. The heat is lost to the surrounding air or ground. The temperature rises until the rate of dissipation from the insulation is equal to the rate of heat production in the conductor. Unfortunately good electrical insulators are often good heat insulators and it is quite possible to produce heat in the conductor at such a rate that the temperature at which equilibrium would be reached would damage the insulation. In addition, the leakage current through the insulation must be considered. This produces heat in addition to that produced by the current in the conductor. For example, a cable operating at 240 V which has an insulation resistance of 240 megohms will allow $240/240 \times 10^6$ amperes to flow and the power loss will be 0.24 mW.

Figure 5.23 shows that as the temperature of a cable insulating material rises its insulation resistance falls. Hence, at a higher temperature the leakage current is greater so that more heat is produced in the insulation. An unstable condition can be created where more heat increases the temperature which lowers the resistance of the insulation allowing more leakage current to flow which produces more heat etc. The cable eventually fails by burning under these circumstances. (See also Loss angle in this chapter.)

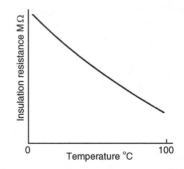

Figure 5.23

PVC becomes soft at temperatures in excess of 80 °C and the conductors tend to migrate through the insulation eventually touching each other or the earth conductor. An upper working temperature limit of 70 °C for PVC is set for this reason.

Permitted volt drops in cables

The 16th edition of the IEE Regulations, regulation 525-01-02, states that the size of the conductor feeding a load shall be such that the volt drop at the load does not exceed 4 per cent of the nominal voltage, not taking into account fluctuations or current surges due to motor starting and the like.

The following data is taken from Tables 4D1A and 4D1B of the 16th edition of the IEE Regulations.

Current-carrying capacities and associated volt drops for unsheathed single-core PVC insulated cables in conduit on a wall.

Conductor cross-sectional area	Single-phase circuit (two cables)	
mm^2	Current-carrying capacity (A)	Volt drop per ampere per metre (mV)
1.0	13.5	44
1.5	17.5	29
2.5	24	18
4	32	11
6	41	7.3
10	57	4.4
16	76	2.7

Correction factors for ambient temperatures when using fuses to BS 88 or BS 1361 (HRC fuses) (Regulation 523-01 and Appendix)

Ambient temperature (°C)	25	35	40	45	50	55
Factor	1.03	0.94	0.87	0.79	0.71	0.61

Effect of ambient temperature on rating

The amount of heat which can be conducted away from a cable depends on the temperature difference between the cable and its surrounding medium. A conductor which is at 60°C in an environment which is at 60°C cannot dissipate any heat. The same conductor in an environment at 5°C will dissipate a large amount of heat.

The upper working temperature of the insulation is fixed at a value at which it will not deteriorate rapidly and at which its mechanical properties will be unimpaired. In the previous section we saw that this was 70 °C for PVC. The heat which can be dissipated therefore depends on the temperature of the surrounding medium, or *ambient temperature*. For a particular current, the heat to be dissipated can be decreased by decreasing the resistance of the conductors. This will involve increasing their cross-sectional areas.

A cable with a particular cross-section which can carry 20 amperes with an ambient temperature of 25°C might need to be replaced with one which has twice the original cross-section to carry this current in an ambient temperature of 50°C.

The increase in area depends not only upon the ambient temperature but on the type of circuit protection employed. Where this is a high-rupturing-capacity (HRC) fuse, the increase is less than when a rewireable semi-enclosed fuse is used. The difference is due to the degree of overloading which the different devices allow before operating. (We will consider HRC fuse protection only, see table of values in the previous section.)

Example 5.1

A 240 V, single-phase circuit has a full-load current of 15 A and is protected by a 15 A, HRC fuse to BS 88. The circuit conforms to Tables 4D1A/B. The circuit runs in a region with ambient temperature 45 °C. Determine the required cross-sectional area of the cable conductors. Calculate the volt drop in the circuit if the run is 20 m in length. Is this satisfactory? (Reg 525-01-02)

From the table the correction factor

for ambient temperature = 0.79.

The cable has to be rated at $\dfrac{15}{0.79} = 19\ A$.

(Note that although the cable must be capable of carrying 19 A, it will carry only the full-load current of 15 A. We must use this value in our volt drop calculations.)

From Tables 4D1A/B we see that a 1.5 mm^2 cable will carry 17.5 A which is not quite enough. It will therefore be necessary to use the next size which is 2.5 mm^2 in cross-section.

The allowable volt drop is 4% of 240 V which is 9.6 V.

The volt drop for 2.5 mm^2 cable is 18 mV/A per m. With 15 A this is 15 × 18 mV = 0.27 V/m.

With a length of 20 m the total volt drop will be 20 × 0.27 = **5.4 V**.

This is less than 9.6 V and so is regarded as satisfactory.

Cables in ducts or in close proximity

Cables which are in ducts or are laid close to each other suffer mutual heating; one cable, becoming warm, heats its neighbours. Where cables touch each other, the number of free paths by which heat can escape to the environment is reduced. The general mass of cables may raise the temperature of the surrounding air or ground. For these reasons it is necessary to reduce the heat generated by the cables and this is done by increasing the cross-sectional area, as already discussed under 'Effect of ambient temperature on rating'. A factor is applied which may be determined by experiment or obtained from sources such as the IEE Regulations or Electrical Research Association data.

A small part of Table 4B1 from the 16th edition of the IEE Regulations is quoted here.

Correction factors for groups of more than one circuit of single-core cables, or more than one multi-core cable 'enclosed' or 'clipped direct' conditions.

	Correction factor							
Arrangement of cables	*Number of circuits or multicore cables*							
	2	3	4	6	8	10	12	14
Enclosed in conduit or trunking, or bunched and clipped direct	0.8	0.7	0.65	0.57	0.52	0.48	0.45	0.43

Select suitable conductors from Tables 4D1A/B for the following circuits. The currents quoted refer to the rating of the HRC fuse protecting the circuit.

(a) 15 A, circuit length 10 m, nominal circuit voltage 250 V. No temperature consideration required.
(b) 15 A, circuit length 10 m, nominal circuit voltage 100 V. No temperature consideration required.
(c) 30 A, circuit length 25 m, nominal circuit voltage 200 V. Ambient temperature 35°C.
(d) 10 A, circuit length 5 m, nominal supply voltage 240 V. Ambient temperature 50°C.
(e) With details as in (d) above but with four identical circuits contained in a conduit.

Example 5.2

Estimate the required cross-sectional area of the conductors to Tables 4D1A/B when they are protected by HRC fuses with a rating of 10 A for the following circuits:

(a) a single circuit situated in an ambient temperature of 25°C
(b) six circuits run together inside trunking, the ambient temperature being 25°C
(c) six circuits run together, the ambient temperature being 50°C.

(a) The ambient temperature correction for 25°C = 1.03.
The cable must be capable of carrying $\dfrac{10}{1.03}$
$$= 9.71 \text{ A.}$$
Using Tables 4D1A/B we see that a 1 mm^2 conductor has a current-carrying capacity of 13.5 A so that this will be adequate.

(b) The grouping factor for six circuits = 0.57. The ambient temperature correction = 1.03.
The cable must be capable of carrying $\dfrac{10}{1.03 \times 0.57}$
$$= 17.03 \text{ A.}$$
This will require the use of 1.5 mm^2 conductor which can carry 17.5 A.

(c) The grouping factor = 0.57. The ambient temperature correction = 0.71 for 50°C.
The cable must be capable of carrying $\dfrac{10}{0.57 \times 0.71}$
$$= 24.7 \text{ A.}$$
This will require the use of 4 mm^2 conductor which can carry up to 32 A.

Example 5.3

A single phase circuit is protected by an HRC fuse to BS 88 with a rating of 20 A. The nominal circuit voltage is 230 V. The load is 30 m from the supply point and fuse. Using Tables 4D1A/B determine the cross-sectional area of a suitable conductor given that there are no temperature or grouping factors to consider.
A 2.5 mm^2 conductor is rated at 24 A and would seem appropriate on first consideration. Investigate the volt drop along its length.
We are allowed 4% of 230 V = 9.2 V.
The 2.5 mm^2 circuit has a volt drop of 18 mV/A/m
Total volt drop = 20 A × 18 mV × 30 m = 10.8 V. This is too great and the conductor size must be increased.
Try 4 mm^2 conductor. Volt drop = 11 mV/A/m.
Volt drop in this circuit = 20 A × 11 mV × 30 m = 6.6 V.

This is less than 9.2 V and is therefore satisfactory. We have in fact used cable rated at 32 A to carry 20 A in order to satisfy the volt drop requirements.

Activity 5.3

We have been using a limited amount of information concerned with PVC-insulated cables in a conduit on a wall. For other cables in other situations other ratings and factors apply. The information may be seen in the complete IEE Regulations or in one of a number of guides to these regulations or in the IEE On-site Guide. From one of these publications obtain data which enables you to compare the performance of various types of cable according to insulation and method of installation and how the rating and temperature factor varies with the type of protection. Why would it be almost certainly better to use MICS cable in a boiler house whereas the choice might be for PVC-insulated cables in conduit in a warehouse? Why use conduit at all when the PVC-insulated cables could be clipped directly to a surface?

Why is it permissible to use 2.5 mm^2 cable in a house ring main protected by a 30 A fuse when this cable is rated at 24 A without applying any factors? Why can a number of industrial 16 A sockets on a radial feeder be supplied through a 2.5 mm^2 cable protected by a 20 A circuit breaker?

Heat sinks

The characteristics of semi-conducting devices and rectifiers and transistors change considerably with a rise in temperature, so much so that the circuit in which they are connected may cease to operate in the manner intended. For example, a rectifier will become conducting in both directions if it is made hot enough. Often a device is permanently damaged and the original characteristics are not re-established by cooling. Where excessive currents have been drawn due to these changes, other components in the circuit may have been damaged.

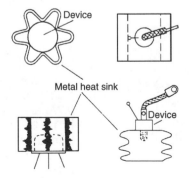

Figure 5.24

Insulation deteriorates at high temperature and the printed circuit board itself or connecting wires can suffer damage. With plug-in components the contacts may suffer since contact springs can become soft and ineffective so introducing a high resistance and more heat into the circuit.

Heat sinks are metal clip or screw-on additions to a device which effectively increase surface area and so aid cooling.

Two types of heat sink are as shown in Figure 5.24. These may be fan cooled in extreme cases.

The insulation of overhead lines

The material most commonly used for overhead line insulators is porcelain. The insulators are shaped from the raw material (a mixture of clay, finely ground feldspar and silica in water) are dried, dipped in liquid glaze and then fired at very high temperature. The glaze forms a glass-like coating providing a surface to which dirt cannot readily stick. It also improves the strength of the insulator so that fracture is more difficult.

One disadvantage of porcelain is that when the glaze is chipped by a power arc or by missiles projected by vandals, water can soak into the body of the insulator. This causes it to become conducting and an electrical discharge takes place which trips out the line while the heat produced dries out the insulator. Such a fault is immensely difficult to find, the line trippling out each time there is rain.

Glass is an alternative material. Toughened glass insulators have a higher breakdown strength under electrical stress than porcelain and if damaged shatter completely rather like the windscreen of a motor vehicle when it is hit by a stone. Missing insulators leave gaps which are easily spotted during one of the regular inspections which are made of the line. The construction is such that although the insulator shatters, the line cannot fall down.

The glass is toughened by heating it to the softening point and then cooling it fairly rapidly. This causes the surface to become hard whilst the centre is still plastic. As the centre cools it tries to contract so pulling in on the outer layers. Before the insulator can fracture, these internal stresses have to be overcome and the force can be six times that required to break ordinary glass.

Suspension insulators can be made of porcelain or glass whilst pin insulators are almost invariably made of porcelain. (See Chapter 1, Figure 1.4.)

Insulators must withstand mechanical and electrical stresses. Heavy lines must be held up off the ground whilst the electrical potential is considerably above that of earth. Even in wet weather the insulator must function and for this reason it has sheds or skirts to keep at least part of the surface dry in almost any weather conditions.

Figure 5.25 shows a pin type insulator supported at the bottom with the line at the top. The steel pin is at earth potential. An electrical breakdown can occur in one of three ways:

1 A dry flashover can occur. This means that an arc will form round the insulator from line to pin along route 1 in Figure 5.25.

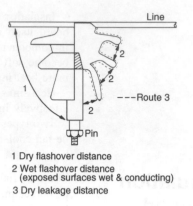

1 Dry flashover distance
2 Wet flashover distance
  (exposed surfaces wet & conducting)
3 Dry leakage distance

Figure 5.25

2   A wet flashover can occur. When the top surfaces of the insulator are wetted by rain they become conducting so that the sheds are disposed to keep the undersides dry. A flashover can occur along route 2 in Figure 5.25.

3   Current can leak from the line over the surface of the insulator along the dotted route 3 in Figure 5.25.

Atmospheric pollution causes dirt to accumulate on the surface of the insulator making it partly conducting and a flashover more likely. Fog wets the insulator overall causing surface leakage to occur especially when smoke and sulphur oxides are also present. Sea spray carried by the wind also has the same effect.

Eddy current and hysteresis losses

When a conducting material is situated in a region of changing magnetic flux a voltage is induced in that material and a current flows. By Lenz's law the direction of this current is such as to create an opposing magnetic flux. Figure 5.26 shows a single turn wound on a conducting core. The current in the winding is increasing. The core flux is therefore increasing and voltages are induced in (a) the current-carrying coil itself, this voltage restricting the rate of rise of the current, (b) the core material, this voltage driving what are called eddy currents which heat the core since it has resistance, (c) any other coil or conducting material in close proximity to the core, this being the basis of the transformer.

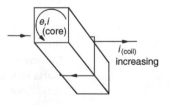

Figure 5.26

The induced voltage in case (a) is a self-induced voltage, the coil possessing self-inductance whilst in cases (b) and (c) we have mutually induced voltages, there being mutual inductance between the coil and core and any other conductor in the vicinity.

From Faraday we know that the induced voltage

$$e = \frac{\text{Change in flux linkages}}{\text{Time taken for the change}}$$

Using symbols: $e = \dfrac{\Delta\phi}{\Delta t}$ the Greek Δ (delta) being read as 'the change in'.

Considering the core as a single turn, when an alternating current flows in the coil in Figure 5.26, the time interval for a current to change from zero to a maximum or from a maximum to zero etc., with the corresponding changes in core flux, is a function of the frequency. The greater the frequency, the smaller the time interval for the change.

Hence, the time interval $\Delta t \propto \dfrac{1}{f}$

So that $e \propto \dfrac{\Delta\phi}{1/f}$ $e \propto \Delta\phi f$ volts

Now since $i = \dfrac{e}{R}$ and power $= ei$ watts

power $= e \times \dfrac{e}{R} = \dfrac{e^2}{R}$ watts

Where $R =$ resistance offered to the current flowing in the core material.

The eddy current power loss $\propto \dfrac{(\Delta\phi f)^2}{R}$

In addition when ferrous cores are used there is a hysteresis loss which is due to the reversal of the magnetic field in the core material. Energy is required to orientate very small groups of atoms called Domains so as to create a magnetic field in one direction during a positive-going half cycle and then to re-orientate them with opposite polarity during the negative-going half cycle.

The power loss is a function of the maximum working flux density (how many of the domains need to be oriented to give the required field strength) and of the number of times this is carried out per second (the frequency).

Hysteresis loss $\alpha\Delta B \times f$ watts.

Magnetic materials

Certain materials such as iron, nickel and cobalt can be made very strongly magnetic, either permanently when they are known as permanent magnets, or temporarily when surrounded by a current-carrying coil, when they are called electro-magnets.

In Figure 5.27 a coil of N turns is shown carrying a current of I amperes. The product of $N \times I$ is known as the magneto-motive force and it is this m.m.f. which is creating the magnetic field which links with the coil. The m.m.f. divided by the length of the coil l is called the magnetizing force which has the symbol H.

$$H = \frac{NI}{l} \text{ amperes per metre}$$

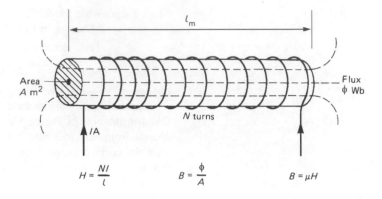

Figure 5.27

This magnetizing force sets up a magnetic field of value ϕ webers. The density of the field is expressed as webers per square metre, symbol B.

$$B = \frac{\phi}{A} \text{ webers per square metre or tesla (T)}$$

The relationship between B and H is called permeability, symbol μ, such that

$$B = \mu H$$

A material which has a high permeability will have a strong magnetic field created in response to a small amount of excitation, that is to say, for a small product of NI. Figure 5.28 shows a set of curves of flux density plotted against magnetizing force for different materials. Observe that the magnetic field strength increases rapidly at first but eventually levels off at values between 1.1 and 1.5 tesla. Where the curves level off, it is said that the materials are saturated. Generally we are interested in the region of the curves before saturation occurs, since here there is maximum magnetic field effect for the minimum energy input to the exciting coil. This means that in this region the permeability is large. If we move into saturation, the magnetizing force H goes on increasing for little or no increase in flux density. This gives low values of permeability.

For example, in a particular iron sample it takes a value of $H = 200$ A/m to set up a flux density $B = 1$ tesla.

Since $B = \mu H$, $\quad 1 = \mu \times 200, \quad \mu = 1/200 = 5 \times 10^{-3}$.

Moving into saturation, it requires $H = 1500$ to set up a flux density $B = 1.3$ tesla.

$$1.3 = \mu \times 1300 \quad \mu = 1.3/1500 = 0.866 \times 10^{-3}$$

Let us now examine the properties of some materials used in devices in which magnetic fields are required.

1 *Ferrite.* The original 'lodestone' was used by mariners who first discovered that this material would always point in a particular direction. A lump of this material, the earliest form of compass, enabled them to navigate. Lodestone is magnetite

---

Test your knowledge 5.6

The permeability of air = $4\pi \times 10^{-7}$ H/m. Determine the value of flux density and total flux within a solenoid 10 cm in length which is wound with 1000 turns carrying 50 A. The cross-sectional area of the air core of the solenoid is 2 cm².

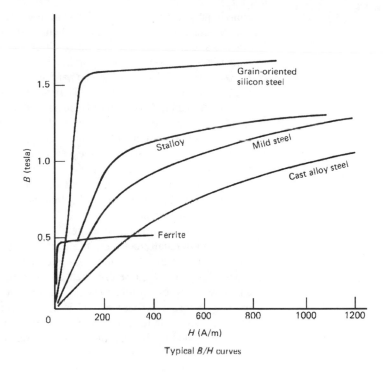

Typical *B/H* curves

Figure 5.28

or ferrous ferrite. These are mixtures of iron oxides. (Rust is an iron oxide as is the black scale which forms when iron is heated.) Ferrites are manufactured at the present time from iron oxides together with oxides of manganese, nickel and zinc. Ferrites have high permeabilities but can only work up to about one half of the maximum flux density of steel. One advantage over steel in certain circumstances is that the body of the material has a very high resistance to electrical currents flowing in it so that eddy current losses will be very small. Some of the ferrites can be permanently magnetized, when they become very difficult to de-magnetize.

2 *Soft iron.* This is very nearly pure iron. It is very soft and can be readily bent or formed into any required shape. It may contain a little silicon (up to about 4 per cent) to increase its resistance to internal current flow. If silicon iron is rolled out from a fairly thick sheet to a much thinner sheet whilst cold and then heated up and slowly cooled to anneal it, it becomes what is known as 'grain oriented'. When used as an electro-magnet, provided that the magnetic field runs in the same direction as the sheet was rolled, it exhibits greater permeability than ordinary soft iron. This, together with its high resistance makes it ideal for use as transformer core laminations.

3 *Soft alloys.* Alloys of nickel and pure iron with possibly small additions of molybdenum and copper are very soft and are known under their trade names, amongst which are Stalloy and Mumetal. They have a permeability which lies between that for nearly pure iron and the grain-oriented variety. Since they

contain no silicon, their resistances are low. Their high permeability and low resistance make them suitable as screening materials. Mumetal was so named after the Greek letter μ which is the symbol for permeability.

4   *Hard alloys.*   These alloys are so hard as to be unrollable when cold, and cannot be filed or bent. They are cast into blocks from the molten material and can only be shaped using a high-speed grinding wheel. Such materials would be found in tools such as files, some spanners and screwdrivers etc. Particularly when used as a magnetic material, they will be made from various mixtures of iron, nickel, cobalt, chromium copper and tungsten. Permeability is fairly high but is not a prime requirement. What is required is the ability to retain their permanent magnetism in any adverse conditions.

5   *Magnetic dust.*   Particles of nickel, iron, molybdenum or pure iron are individually covered in an insulating material, mixed with an insulating resin and then pressed into any required shape. They may then be operated as electro-magnets without any possibility of current flowing through the material from particle to particle. These materials have lower permeabilities than iron since there is less active material present (a lot of space is taken up by the filler). They are, however, almost immune from eddy current losses.

6   *Air.*   Although not generally considered to be a magnetic material, there are a number of applications where an electromagnet with an air core is the only possibility. It has very low permeability ($4\pi \times 10^{-7}$), but suffers no internal losses.

Minimization of losses

In electrical machines, working at frequencies up to a few kiloherz, high values of flux density are required so that a small cross-sectional area of magnetic material may be used to give the required value of flux.

Flux density × area of magnetic core = total flux

Using a small area minimizes the weight and hence the cost of the equipment.

The eddy current losses in the core are reduced by increasing its resistance. This is achieved by the use of laminations. A lamination is a very thin sheet of material, less than 0.4 mm thick, which is carefully cleaned and varnished or anodized on one side. Many laminations are pressed together to form the required cross-section of the core. An alternative, but less effective method uses thin sheets of paper between the laminations. Figure 5.29 shows a simple arrangement. Eddy currents can no longer flow as shown in Figure 5.26 but are constrained to very small loops within each of the insulated laminations. The use of 4 per cent silicon steel increases the resistance of the individual laminations while the magnetic strength may be increased for a given coil and current by the use of grain-oriented steel.

Laminations of soft nickel-iron alloys at thicknesses of less than 0.1 mm may be used at frequencies up to about 10 MHz. For frequencies higher than this, since the eddy current losses are

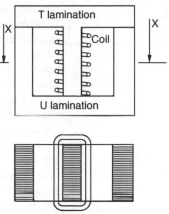

Figure 5.29

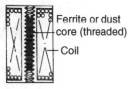

Figure 5.30

proportional to the square of the frequency these may be considerable even with the thinnest laminations. Thus dust or ferrite cores are used up to about 150 MHz in communications equipment. In ultra-high-frequency applications, air cores may have to be resorted to since air suffers no hysteresis or eddy current losses. Figure 5.30 shows a small coil with a ferrite or dust core which is threaded. It may be screwed into or out of the coil to adjust the inductance value.

Permanent magnets

Permanent magnets are made from hard alloys. They are heated to a high temperature, shaped and allowed to cool inside a coil carrying a large direct current. They then become permanently magnetized. They are used, for example, in permanent magnet moving-coil DC instruments, loud speakers and small electric motors. They may become demagnetized if they are heated to a high temperature either by an external heat source or if they are operated in the presence of a high-power alternating magnetic field from another piece of equipment, when the alternating field will cause eddy currents to flow within the permanent magnet (as already described under 'Eddy current and hysteresis losses'). This will produce internal heating and possible demagnetization. One solution to this problem is the use of permanent magnet ferrites. These are not so strong as the metal types but do not suffer eddy current losses so do not heat up in the presence of external alternating fields.

Magnetic screening

If a coil carrying alternating current is enclosed within a cylinder (a screen) of soft alloy (Mumetal) it becomes what is in effect a transformer with the screen as the secondary, the coil being the primary. By Lenz's law, the current flowing in the screen will produce a magnetic field in opposition to that produced by the coil. The two magnetic fields cancel and there is no magnetic effect outside the screen. (See 'Conductor screening and coaxial cable' and Figure 5.13.) Since Mumetal has a high permeability, the magnetic field is strong and hence the eddy currents are large even though the cross-sectional area of the metal is small. As the frequency increases, the e.m.f. induced in the screen increases and the eddy currents increase in magnitude. The screen therefore becomes more and more effective as the frequency increases. Above 20 kHz the Mumetal screen may be replaced by one of copper or aluminium since although their permeabilities are less, the increasing frequency produces the necessary increase in eddy currents for good screening.

Activity 5.4

In this chapter we have looked briefly at the properties of permanent magnets and high-permeability soft materials. Investigate these materials more fully using the available manufacturers, data and library material.

Compare the properties of materials suitable for (a) the core of a power transformer, (b) the core of a transformer or choke working in the megahertz range, (c) use as a permanent magnet operating in a region of a strong alternating magnetic field and (d) a screen for a coil carrying current (i) at fairly low frequencies, (ii) in the hundreds of kilohertz range.

Problems on materials

1   Although silver is the best electrical conductor, it is not used in cable manufacture. Why is this?

2   How is the low tensile strength of aluminium compensated for in the construction of overhead lines?

3   Hard drawn copper is not very flexible. How is flexibility built into cables and lines with copper cores?

4   Why would PVC insulation not be used in hot situations? Suggest a suitable alternative type of insulation.

5   Calculate the volt drops in the following circuits using Tables 4D1A/B:

(a) c.s.a. 2.5 mm^2, current 20 A, length of run 3 m
(b) c.s.a. 1.5 mm^2, current 12 A, length of run 5 m
(c) c.s.a. 4 mm^2, current 25 A, length of run 10 m.

[(a) 1.08 V, (b) 1.74 V, (c) 2.75 V]

6   Determine suitable conductor sizes for the following circuits (Tables 4D1A/B). The nominal supply voltage is 240 V.

	Fuse rating (BS 88)	Temperature of run (°C)	Number of circuits in a single conduit	Length of run (m)
(a)	10	35	1	25
(b)	10	40	3	6
(c)	15	50	2	40
(d)	30	45	6	8

[(a) 1.5 mm^2, (b) 1.5 mm^2, (c) 4 mm^2, (d) 16 mm^2]

7   A single-phase circuit carrying 50 A has a power loss of 1 W per metre run due to conductor heating. There is a force between the conductors of 0.1 N/m run due to the magnetic effect. Calculate the power loss and the force between the conductors when a short-circuit occurs and the current rises to 5000 A.

[10 kW, 1000 N both per metre run]

8   A cable draws a capacitive current I_C of 500 mA and a current in phase with the voltage, I_R of 1 mA. What is the value of δ, the loss angle? Why does this need to be as small as possible? If the applied voltage to the cable is 1000 V, what is the power loss?

[0.115°, 1 W]

9   How is the choice of cable size affected by the type of fuse employed to protect the circuit?

10   What property has magnesium carbonate as a cable insulant which no other insulating material can match?

11   What methods are there in cable construction whereby air is used as the principal insulant but the conductors are maintained at their correct separation? Draw simple sketches to illustrate your answer.

12   Name the two principle insulating materials used in overhead line construction.

13   What is the difference between a pin insulator and a suspension insulator?

14   Why is flashover on an overhead line insulator more likely to happen on a rainy or foggy day than on a dry day? What features are built into the insulator to minimize the occurrence of flashover?

15   Why is it necessary to employ conductors with greater cross-sectional area when they are run in groups as compared with running single circuits?

16   A cable has an insulation resistance of 20 MΩ and a capacitance of 0.01 μF. The supply voltage is 66 000 V at

a frequency of 50 Hz. With reference to Figure 5.3 calculate the value of I_R and I_C and hence determine the value of the loss angle δ. What is the power loss in the cable?

[$X_C = 318$ kΩ, $I_R = 3.3$ mA, $I_C = 0.207$ A, $\delta = 0.91°$, Power loss in dielectric 217.8 W]

17 What is the purpose of oil filling under pressure in a paper-insulated power cable?

18 Why is it necessary to use coaxial cable to interconnect equipments operating at very high frequencies?

19 Describe two types of cable for use at radio frequencies.

20 What are the advantages of using printed circuit boards in a television receiver as opposed to individually wired chassis?

21 Describe two processes which are available for the manufacture of printed circuit boards.

22 A printed circuit board has one circuit formed from foil 0.035 mm thick and 0.5 mm wide. The permitted current density for 40°C temperature rise is 50 A/mm^2. Calculate the maximum permissible current in the circuit.

[0.875 A]

23 Calculate the current densities in the following printed circuit board conductors:

(a)   0.07 mm × 1 mm carrying 5 A
(b)   0.106 mm × 2 mm carrying 4 A.

[(a) 71.4 A/mm^2, (b) 18.87 A/mm^2]

24 What is the function of a heat sink?

25 What difficulties might be encountered in situating a permanent magnet next to a coil carrying alternating current?

26 Suggest suitable core materials for the following, giving reasons:

(a)   A power transformer with a rating of 500 MW in service 24 hours every day
(b)   A portable power transformer rated at 500 W for use only a few hours in a week. (*Hint* (a) and (b), think about what importance the losses have in each of the situations
(c)   A coil operating at 100 MHz
(d)   A coil operating at 1000 MHz.

27 In what situation might a permanent magnet ferrite be used in preference to a metal alloy permanent magnet?

28 A coil is to be screened to prevent interference with adjacent circuits. Suggest suitable materials for the screening can for frequencies (a) 15 kHz, (b) 10 MHz.

29   The hysteresis loss in a transformer core is 75 W and the eddy current loss is also 75 W when it is operated at 50 Hz. What would be the expected hysteresis and eddy current losses if the frequency were doubled to 100 Hz provided that the core flux remained unchanged?

[Hysteresis loss 150 W, eddy current loss 300 W]

Multiple choice questions

1   Copper and aluminium are used as electrical conductors in a virtually pure state since the addition of alloying elements:

A   makes them too soft
B   makes them too hard
C   increases their electrical resistance
D   makes it more difficult to machine them or to draw them out into wire.

2   One of the reasons that aluminium is chosen in preference to copper for overhead lines is that, for a given resistance, the aluminium line is:

A   smaller in cross-section than copper
B   more flexible than copper
C   although larger than copper, is lighter
D   the aluminium strands are easier to lay around the steel core than copper strands which are much harder.

3   Lengths of overhead line are joined together using:

A   crimped terminals and bolting
B   compression sleeves
C   split ferrules and solder
D   welding.

4   Three circuits in a conduit operating at 240 V are 25 m in length and protected by 15 A fuses operate in an ambient temperature of 50°C. They would each need to have a cross-sectional area of:

A   1.5 mm²
B   1 mm²
C   2.5 mm²

to satisfy the conditions laid down in the IEE Regulations.

5   At 132 kV, underground cable insulation is almost invariably:

A   paper
B   PVC
C   polythene.

6   For cables which are to operate at 415/240 V in a situation where fire resistance is required, a suitable insulating material is:

A   PVC
B   polychloroprene
C   paper
D   magnesium oxide (mineral)
E   polythene.

7   The magnetic force between conductors may be disruptive under short-circuit conditions. The force may be reduced by:

A   spacing the circuit conductors closer together
B   increasing the conductor spacing
C   using different insulating materials, e.g. replacing PVC with oil-impregnated paper
D   putting the conductors inside a plastic conduit.

8   Coaxial cable is used to interconnect pieces of equipment when:

A   very large currents are involved
B   it is essential that no magnetic field shall exist round the cable
C   the effect of cable heating is to be minimized
D   extra high voltages are to be used.

9   Gold is often used for multi-contact connections at low voltage. This is because:

A   it has very low resistivity
B   it is soft and ductile
C   it gives low contact resistance at low pressures
D   it is easily soldered to connecting wires joining it to another piece of equipment.

10   Pre-stressed glass is used as an insulating material for overhead lines because:

A   when it is damaged it completely shatters
B   it has much better electrical properties than porcelain
C   it may be formed into an insulator with a much larger flashover distance so has better performance in foggy weather
D   it is much better than porcelain when used in pin insulators since it is more flexible and so less likely to fracture when subject to a cross pull (by wind, for example)
E   the manufacturing costs are considerably less.

11   Permanent magnets may be of hard alloy steel or ferrite. Ferrite might be used:

A   where a strong magnetic field is required
B   where it is essential to eliminate hysteresis loss
C   where it is to operate in a region in which there is a strong alternating magnetic field.

12   Dust cores may be used in transformers and coils for operation at frequencies between 100 MHz and 150 MHz. They are used to:

A   minimize hysteresis loss
B   increase the maximum working flux
C   make fabrication of various core shapes easier than when using laminations
D   reduce eddy current losses.

6 Illumination

Summary

This chapter looks at various types of light source based on hot tungsten wire or gaseous discharge and examines their luminous efficiency. The basic laws of illumination are explained. The method of calculating the level of illuminance of a surface using the inverse square law and a polar diagram is demonstrated. The method of using utilization and maintenance factors to establish the number of luminaires to provide uniform illumination is examined. The effects of diffusing and prismatic luminaires are described. The balance between daylighting and artificial illumination is investigated.

White light may be resolved into a band of colours called the spectrum ranging through red, yellow, green, blue and violet by passing it through a glass prism as shown in Figure 6.1. The same effect occurs naturally when sunlight falls on water droplets, forming a rainbow against a background of dark cloud.

These colours are due to electro-magnetic radiations similar to radio waves but of a much higher frequency.

Red light has a frequency of 4×10^{14} Hz
Violet light has a frequency of 7.5×10^{14} Hz

By comparison, radio transmissions have frequencies starting at about 1×10^5 Hz in the long-wave band extending to about 1×10^8 Hz in the very-high-frequency band. Television signals need a higher frequency still, for example 8×10^8 Hz, and in some ways these act similarly to light, it being difficult to receive signals over a hill, for example.

The speed of electromagnetic radiations such as light and radio is 3×10^8 m/s (closely).

At frequencies slightly lower than red there is infra-red radiation which, although invisible, can be detected by the human skin since it is warming. At the other end of the spectrum at frequencies above that of violet light, there is ultra-violet radiation which also has an effect on human skin, turning it brown as a means of protection. Excessive doses of ultra-violet radiation can be dangerous, burning the skin, damaging the eyes and causing skin cancers.

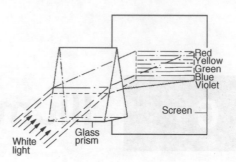

Figure 6.1

The essential source of heat and light for the earth is the sun which provides radiations from the ultra-violet, right through the visible spectrum, and into the infra-red region. Much harmful radiation is absorbed by the earth's atmosphere and never reaches the ground but in areas where this is very clear and possibly where the protective ozone layer has been depleted there are still dangers and barrier creams, suitable clothing and eye protection are used by persons exposed to bright sunlight.

Many solid bodies emit light when raised to a sufficiently high temperature. The ordinary incandescent light bulb, which contains a fine tungsten wire being heated by the passage of an electric current, is a good example. Figure 6.2 shows the various frequencies of radiations in the output from such a lamp. Note the relatively large amount of heat produced (infra-red region) and how little of the radiation is in the visible region.

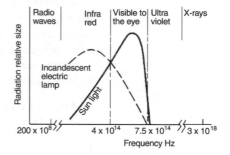

Figure 6.2

Lighting standards

Since measurements of illumination were first made it has been necessary to have a standard with which the various light sources could be compared.

The original standard was the wax candle. The exact level of illumination was difficult to reproduce each time it was required since the grade of wax, the material from which the wick was made and its burning length had to be specified and accurately controlled. The standard was followed by a lamp burning pentane gas and later by a standard incandescent lamp.

In 1948, with the introduction of SI units, the platinum standard was adopted. It has already been noted that when solid bodies are heated they can emit light. The nature of the light depends on the temperature and this is rather difficult to determine accurately. The problem is overcome by using a material which emits light at the temperature at which it solidifies. It therefore exploits the phenomenon that when materials change state, from liquid to solid or liquid to gas, they do so at constant temperature. A common example is that of water as it freezes into ice (temperature constant at 0°C) or evaporates into steam (temperature constant at 100°C).

Platinum melts or freezes at 1773°C and at this temperature emits light since it is white hot. The platinum, contained in a small tube within an insulated box, is melted electrically. During the time it is solidifying the temperature remains constant so that the light intensity at a small viewing hole at the top of the box is constant. This light is used as the standard and other light sources are compared with it. Another reason that platinum is ideal for the standard is that it remains chemically unchanged during successive melts.

Luminance and luminous intensity

The luminance of the viewing hole in the standard is defined as 60 candelas per square centimetre (60 cd/cm^2). Let us examine what this means.

Luminance can be interpreted as a measure of the discomfort caused by looking at a light source. Imagine trying to look directly at the noonday sun. It will cause pain and may indeed damage the eye if there is prolonged exposure. The sun has very high luminance estimated to be about 160 000 cd/cm^2 at noon in summer at Greenwich. In contrast a 1500 mm, 80 W fluorescent tube commonly used for lighting offices and shops has a luminance of about 1 cd/cm^2 and this may be viewed without risk. Imagine now being in a completely darkened room when a small hole is cut in one of the blinds through which the bright sun may be viewed. The luminance of the sun viewed through the hole is very high but how much light actually enters the room through this hole? It is minimal and it would be difficult to see things in the room. If, however, a single fluorescent tube is switched on in the room, far more light is provided than by the sun shining through the small hole.

It can therefore be deduced that light-radiating capability or *luminous intensity* is a product of luminance and area. Thus in the above example the sun with its very high luminance is projecting light from a small area and is providing less light than a fluorescent tube which has a very low luminance but a large surface area.

Since the platinum standard has a luminance defined as 60 cd/cm^2, a hole in the top of the standard having an area of 1/60 cm^2 would result in a light-radiating capability of 1 cd. The symbol for light-radiating capability or luminous intensity is I so that in this case $I = 1$ cd.

Example 6.1

The filament of a 100 W clear incandescent lamp has a luminance of 650 cd/cm^2. If the filament has a projected area of 0.1 cm^2 as shown in Figure 6.3, what is the light-radiating capability in candelas?

I = luminance × area

  = 650 × 0.1

  = 65 cd

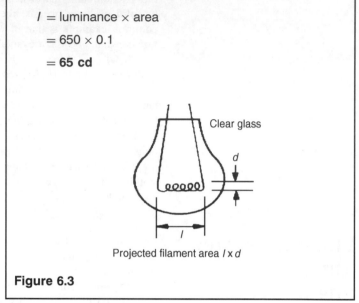

Projected filament area l x d

Figure 6.3

Instead of using clear glass, the interior surface may be treated to create the pearl lamp. This makes the filament appear to have a larger area. If the light-radiating capability is unchanged, the luminance must have decreased. This will make the lamp less uncomfortable to look at and where it may be viewed directly this lamp would be preferred. Alternatively, where the lamp is enclosed in some form of shade or diffuser, making direct viewing impossible, the clear lamp would be quite suitable.

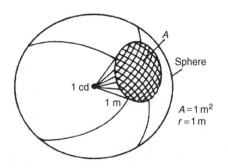

Figure 6.4

The lumen and the steradian

Figure 6.4 shows a light source with uniform luminous intensity of 1 cd at the centre of a sphere of radius 1 m. A uniform source is considered to radiate light equally in all directions but this is

Test your knowledge 6.1

A white screen is 10 m wide and 3 m in height. It is to be illuminated by shining light from a projector on to it. The surface illuminance is to be 50 lux and 85 per cent of the light from the lamp in the projector reaches the screen.

(a) If the lamp in the projector yields 18 lumens for each watt of input, determine a suitable rating for the lamp.
(b) What is the value of light-radiating capability *I* cd for this lamp assuming uniform light-radiating capability?

not achievable absolutely in practice since all lamps have a dark area where the energy is fed in.

If we consider an area *A* on the surface of the sphere of 1 square metre, the solid angle so formed is called a steradian, symbol ω.

The amount of light from the source which passes through the cone, i.e. within the solid angle of 1 steradian, is 1 lumen.

The amount of light is called luminous flux and has the symbol ϕ. Since the radius of the sphere is 1 m, its surface area $= 4\pi r^2 = 4\pi \times 1^2 = 4\pi$ m^2. Now since each square metre of surface forms 1 steradian, 4π square metres must form 4π steradians.

The total solid angle within the sphere $= 4\pi$ steradians

Because there is a luminous flux of 1 lumen in each steradian from a 1 cd source, the total light flux from such a source must be 4π lumens. In Figure 6.4, 1 lumen is falling on 1 m^2 of area and this gives a level of surface illumination or *illuminance* of 1 lux.

Illuminance has the symbol *E*. Hence $E = 1$ lux.

To sum up: 1 cd provides 1 lumen through each steradian,
I cd provides *I* lumens through each steradian
and $I \times \omega$ lumens through ω radians

The total number of lumens from a uniform source is therefore $4\pi I$ lumens ($\omega = 4\pi$).

Example 6.2

A lamp has a uniform luminous intensity of 100 cd. What is the total light output of the lamp in lumens? If all of this light is projected on to an area of 10 m^2, what is the illuminance of that area?

Total luminous flux $= 100 \times 4\pi$

$= 1256.7$ lm

On an area of 10 m this results in an illuminance of $\frac{1256.7}{10}$ lumens on each square metre.

Since the number of lumens falling on 1 m^2 has been defined as the illuminance in lux (lx),

$E = \textbf{125.67 lx}$

The inverse square law

In Figure 6.5 the source of illumination has a luminous intensity of *I* candela. The level of illumination at surface A_1 is due to *I* lumens on 1 square metre since the solid angle is 1 steradian.

$E = I$ lux

The light rays are all meant to be at right angles to the surface being illuminated. If the surface A_1 is removed so that all the light rays which illuminated it now travel two metres to illuminate surface A_2, the luminous flux has now spread over four square

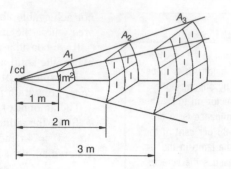

Figure 6.5

metres. There are now I lumens on 4 square metres.

$$E = \frac{I}{4} \text{ lumens per metre}^2 \text{ or lux}$$

For surface A_3 the same luminous flux is distributed over nine square metres.

$$E = \frac{I}{9} \text{ lux}$$

Generally then, the illuminance $E = I/d^2$ lux where d is the distance of the illuminated surface from the source in metres.

The cosine law

In the previous section the light rays were all normal to the surface being illuminated. In the case of street lighting, for example, this could be true only for the position immediately below each lamp.

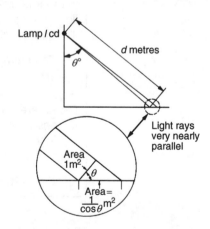

Figure 6.6

Consider a single lamp and a point on the road surface some distance from the base of the lamp as shown in Figure 6.6. At this range the rays of light are considered to be very nearly parallel. The level of illumination on the area of one square metre shown normal to the light rays is I/d^2 lux. As this area is tilted through the angle $\theta°$ so that it lies in the horizontal plane, the same luminous flux now illuminates an area of $1/\cos\theta$ metre^2. The level of illumination on this increased area now becomes

$$\frac{I}{d^2} \div \frac{1}{\cos\theta} = \frac{I\cos\theta}{d^2} \text{ lux}$$

Example 6.3

Calculate the level of illumination (illuminance) on a horizontal surface due to a single lamp with uniform light-radiating capability of 2000 cd mounted 10 m above the surface at points:

(a)   immediately below the lamp,
(b)   10 m from this point on the horizontal surface,
(c)   20 m from the original point on the horizontal surface.

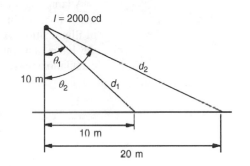

Figure 6.7

(a)   Since the lamp has uniform light radiating capability, the value 2000 cd may be used throughout. (See Figure 6.7.)
      Immediately below the lamp the light rays fall normally on the surface.

$$d = 10 \text{ m} \qquad E = \frac{2000}{10^2} = \textbf{20 lux}$$

(b) $\tan\theta_1 = 10/10 = 1$. Thus $\theta_1 = 45°$. This may be obtained by scale drawing if preferred.
      By Pythagoras' theorem:

$$d_1^2 = 10^2 + 10^2 = 200$$

$$E = \frac{2000}{200}\cos 45°$$

$$= 10 \times 0.707$$

$$= \textbf{7.07 lux}$$

(c) $\tan\theta_2 = 20/10 = 2$. Thus $\theta_2 = 63.5°$.

$$d_2^2 = 10^2 + 20^2 = 500$$

$$E = \frac{2000}{500}\cos 63.5° = \textbf{1.784 lux}$$

Test your knowledge 6.2

A light fitting with uniform light-radiating capability is mounted 12.5 m above level ground. At a point on the ground 20 m from the base of the lamp standard the illuminance is measured and found to be 2 lx. Determine:

(a)   the light-radiating capability of the lamp in candelas

(b)   the total lumen output of the lamp.

The polar diagram

A lamp without any means of directing the light produced radiates a great deal of its light in directions other than those required, up into the sky in the case of a lamp situated outdoors, for example. If a reflector or some form of shade or lantern is used in conjunction with the lamp, light can be radiated efficiently in those directions required.

The light-radiating capability of the fitting is now no longer uniform and in order to calculate the illuminance of a surface it will be necessary to use the value of luminous intensity in the direction towards the surface, e.g. along the lines inclined at θ_1 and θ_2 to the vertical in Figure 6.7. A polar diagram provides information on the luminous intensities in particular directions.

Polar diagrams for particular lamps and reflectors may be obtained from the lamp manufacturers. Alternatively they can be obtained experimentally by using the following procedure.

A lamp, optionally with reflector, is mounted in a room or large box with matt black walls and in which there is no other source of illumination. A light meter is situated at a known distance from the lamp filament. This distance is often one metre to make the calculations easy. The arrangement is shown in Figure 6.8.

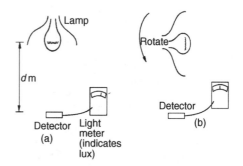

Figure 6.8

The light meter measures illuminance due to the lamp at the light-sensitive surface of the detector.

$$\text{Illuminance } E = \frac{I}{d^2} \text{ lux}$$

If the distance involved is one metre then the meter reading in lux is numerically equal to the luminous intensity I of the source. In Figure 6.8(a) the luminous intensity in a direction along the axis of the lamp is being found.

If the lamp is turned through $90°$ as shown in Figure 6.8(b) the luminous intensity of the lamp at right angles to its axis is being found. The process is carried out at intervals of $10°$ or $20°$ throughout a complete revolution of the fitting. Thus the luminous intensity of the lamp and its reflector in any particular direction is evaluated. A plot of the results on polar graph paper is referred to as the polar diagram.

The light-radiating capability of the fitting is scaled from the origin in the direction required. For instance, directly downwards

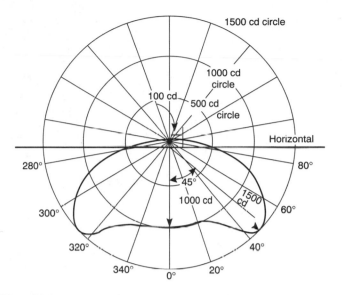

Figure 6.9

in Figure 6.9 it is 1000 candelas whilst in the horizontal direction it is only 100 candelas. At 45° to the horizontal it is 1500 cd.

The light-radiating capability above the horizontal is virtually zero since this particular fitting is designed to project most of the light produced on to the floor.

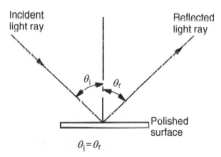

Figure 6.10

Reflection and refraction of light

Light may be directed on to the required area by reflection or refraction. Reflection of light takes place to some extent at most surfaces. If the surface is highly polished, as with a mirror, the reflection is said to be 'specular'. The angle of incidence is equal to the angle of reflection as shown in Figure 6.10. Specular reflection is used in the parabolic reflector which can be found in floodlighting fittings or luminaires, in cinema projectors, and searchlights etc. (Figure 6.11).

The white enamelled steel industrial reflector reflects most of the light downwards as shown in Figure 6.12, distributing it over a wide area. Where the surface is matt finished, diffused reflection takes place. This gives a softer lighting effect although it is not so efficient as specular reflection due to light absorption by the surface. The colour of the surface also affects reflection, light colours obviously reflecting more light than dark ones.

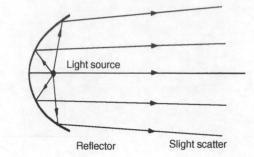

Figure 6.11

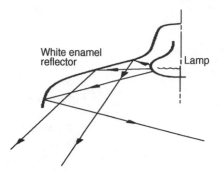

Figure 6.12

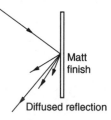

Figure 6.13

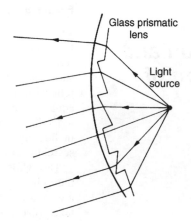

Figure 6.14

Refraction or bending of light rays takes place when they pass through a prism as shown in Figure 6.1. The refracting luminaire is used for street lighting and in motor vehicle headlamps. An array of prisms may be seen on the inside of the glass as shown in cross-section in Figure 6.14. The vehicle headlamp uses a combination of reflection (Figure 6.11) and refraction (Figure 6.14) to direct the beam and give sharp top cut-off to minimize dazzle to oncoming traffic. By a combination of these methods a polar diagram of almost any shape can be produced enabling light to be projected in the direction required. If the object of a luminaire is to diffuse the light so as to reduce the luminance of the source, then diffusing glass globes may be used. These will be finished pearl or opal on the inside or again may make use of much smaller prisms which have the effect of scattering the light as shown in Figure 6.15.

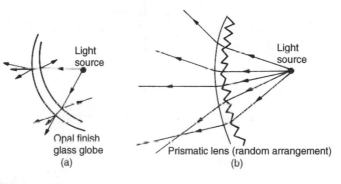

Figure 6.15

Use of the polar diagram

Suppose the surface level of illumination or illuminance provided by a lamp in a specific type of luminaire is required. The polar curve is plotted or obtained from the manufacturer. A typical example is shown in Figure 6.16.

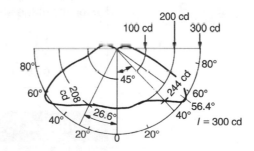

Figure 6.16

The following example illustrates the method to be adopted.

Example 6.4

A single lamp in a reflector has a polar diagram as shown in Figure 6.16. It is mounted on a lamp standard at a height of 10 m above a horizontal flat plane. Calculate the illuminance on the plane:

(a) immediately below the lamp
(b) 10 m from the base of the lamp standard.

The first thing is to draw the arrangement of lamp, standard, and surface to be illuminated as in Figure 6.17. The lamp is considered to be situated at the origin of the polar diagram. The light illuminating the spot immediately below the lamp is shining directly downwards through zero degrees on the polar diagram. By direct reading from the curve, the luminous intensity in this direction is 200 cd.

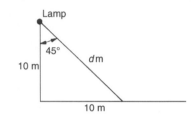

Figure 6.17

(a) $E = \dfrac{I}{d^2} \cos \theta$ lux and in this case $\theta = 0°$

  $E = \dfrac{200}{10^2} = \mathbf{2\ lux}$

(b) At a distance of 10 m from the base of the standard, the light is projected at an angle of 45° from the vertical.

  $d^2 = 10^2 + 10^2 = 200$

From the polar diagram, at 45° to the vertical the luminous intensity of the source is 244 cd.

  $E = \dfrac{244}{200} \cos 45°$ according to the cosine rule

  $= \mathbf{0.863\ lux}$

Activity 6.1

Plot a polar diagram for a lamp and its fitting and use this to estimate the illuminance on a surface at a specific distance from the fitting.

Using a lamp and light fitting of your choice proceed as described in the above section to obtain the necessary readings to plot its polar diagram. Repeat the procedure for this lamp without any form of shade or reflector. Compare the shapes of the curves and discuss the usefulness and applications for the shade/reflector used.

Use your polar diagrams to calculate the expected surface illuminance at various distances and attitudes from the fitting of your choice. Discuss the changes which are brought about by the inclusion of the light fitting as opposed to using the bare lamp. What other advantages are there to using a light fitting as opposed to a bare lamp? Discuss the problems of light pollution now prevalent in built-up areas. With a fitting mounted on a lamp standard, what shape of polar diagram would tend to give near constant level of surface illumination as distances from the base of the lamp standard increase?

Obtain makers' literature relevant to light fittings and their uses.

What type of polar distribution would you be looking for when designing (a) general illumination of a factory working space, (b) spot lighting for the exterior of a building and (c) lighting on an urban road where the standards are 4 m in height. Discuss the reasons for your choices.

Example 6.5

Consider the case of two lamps 20 m apart as shown in Figure 6.18.

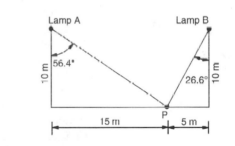

Figure 6.18

The mounting heights and polar diagram for each lamp are as in Example 6.4. Calculate the level of illumination on a line joining the bases of the two standards at a distance of 15 m from one of them. This is point P in Figure 6.18. The illumination at point P comes from two sources. Consider them separately.

From lamp A: $E = \dfrac{300(\text{from Figure 6.16})}{10^2 + 15^2} \times \cos 56.4°$

$= 0.51$ lux

From lamp B: $E = \dfrac{208}{10^2 + 5^2} \cos 26.6° = 1.48$ lux

Total illuminance from both sources $= 1.48 + 0.51$
$$= \mathbf{1.99 \ lux}$$

Activity 6.2

We have seen how the use of lanterns can help to give a desired distribution of light. In Example 6.4, the illuminance under the lamp standard was found to be 2 lux and when considering two lamps in Example 6.5 we found that the illuminance between the standards at point P was closely 2 lux. The level of illuminance between standards is influenced by the polar distribution and the spacing between successive lamps. Situating lamps very close together is likely to result in a more uniform distribution than when they are separated by very large distances but the overall effect will depend on mounting height.

Investigate these effects. Take a light meter on the street and take readings of illuminance due to street lighting comparing, say, a trunk road where the luminaires are at higher level and relatively close together and an urban street on a housing estate where the standards will be shorter and most likely further apart or if not, then using much smaller light sources.

What arrangement of luminaires is there at a sports complex or football park? How uniform is the illuminance across such an area?

Test your knowledge 6.4

The security lamp on a house is mounted 2 m above the ground. The gateway to the property is 10 m away. It is desired to have a surface illuminance at the gate of 40 lux. On the polar diagram for the lamp, at what angle to the vertical are the light rays which illuminate the gate area and what is the required value of light-radiating capability in this direction (I cd)?

Indirect illumination of a surface

The previous examples using the polar diagram show how the illumination at a point may be calculated when that illumination is due to a single source, with no reflecting surfaces nearby, e.g. as in road lighting. If there are two or more sources, again with no reflecting surface nearby, then the illuminance present is the sum of the values due to the individual sources. With indoor illumination, however, this method cannot be used since the distances are small and much of the illumination is the result of reflection from ceilings and walls.

Consider a workshop which requires a given illuminance at the bench tops. The benches receive light in any number of ways as shown in Figure 6.19. Some comes directly from the fitting and some by reflection from the walls, ceiling and floor. Note that light can be reflected several times, losing a little by absorption on each occasion.

The amount of light received after reflection depends on the colour and cleanliness of the surfaces involved. It also varies

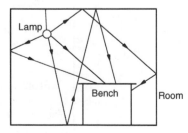

Figure 6.19

according to the height of the ceiling and the shape of the room. The arrangement of the reflecting surfaces, their condition and colour is taken into account by the use of a 'coefficient of utilization'.

The coefficient of utilization is defined as

$$\frac{\text{Luminous flux arriving at the working surface}}{\text{Total luminous flux supplied by the lighting fittings}}$$

This has been determined experimentally by the Illuminating Engineering Society for many different combinations of colour and room proportions and may be found from reference tables in the IES code. The IES also specify the requirement for illuminance at the working surface for various types of work and the table quotes some of these.

Workshops and factories	Illumination recommended in lux
Average for general work	100–150
Assembly – large work	100
Radio assembly	200
Engraving	500
Warehouses	50
Drawing offices	300

Example 6.6

A workshop 30 m × 40 m is to be illuminated using gas-filled tungsten filament lamps in standard dispersive reflectors. They are situated 2.5 m above the working plane on which an illuminance of 100 lux is required.
  The coefficient of utilization is 0.6.
  Calculate the total luminous flux required from the fittings.

100 lux = 100 lumens on each square metre
Area to be illuminated = 30 × 40 m^2
$$= 1200 \text{ m}^2$$

Luminous flux at the working surface = 1200 × 100
$$= 120\,000 \text{ lm}$$

Since the coefficient of utilization

$$= \frac{\text{Luminous flux at the working surface}}{\text{Total input of luminous flux}}$$

$$0.6 = \frac{120\,000}{\text{Input}} \quad \therefore \text{Input} = \frac{120\,000}{0.6} = \textbf{200\,000 lumens}$$

During the life of an electric lamp its light output decreases due to ageing of the filament or of the phosphors employed in fluorescent tubes. Lamp reflectors also become less efficient due to an accumulation of dust. Inevitably some lamps fail altogether. In a factory it is not usually economic to replace single lamps especially if they are rather inaccessible. It is better to wait until either 5 per cent of the installed lamps have gone out or all the lamps have been in operation for their design life before replacing all the lamps at the same time. This can usually be done when the factory is not in production.

Where a minimum level of illumination is specified, allowance is made for the deterioration of lamps by providing a greater level of illumination than this minimum when the installation is new. At the end of the design life of the lamps the output will still be maintained at or above the minimum requirement.

A depreciation factor is used to calculate the original input. If the illuminance in the workshop in the previous Example 6.6 must never fall below 100 lux and a depreciation factor of 1.3 is recommended, then the input when new must be $1.3 \times 200\,000 = 260\,000$ lumens. Alternatively a maintenance factor may be used:

Maintenance factor

$$= \frac{\text{Light output at the end of the design life}}{\text{Light output when new}}$$

Again from the previous example, the maintenance factor would have been 0.77.

$$0.77 = \frac{200\,000}{\text{Light output when new}}$$

$$\text{Light output when new} = \frac{200\,000}{0.77} = 260\,000 \text{ lumens}$$

$$\text{Depreciation factor} = \frac{1}{\text{Maintenance factor}}$$

Clearly whichever factor is used in calculations it must be used to give an original light input which exceeds that at the end of the design life.

It is now necessary to examine the numbers and arrangement of lamps by which the required illuminance may be provided.

Data for gas-filled tungsten filament lamps is as follows:

Power in watts	60	100	200	300	500
Total output, lumens	567	1135	2600	4140	7500

Since a 60 W lamp gives 576 lumens, to provide 260 000 lumens would require $260\,000/576 = 452$ lamps. Alternatively using 500 W lamps, $260\,000/7500 = 35$ lamps would be required.

The effect of having a few large lamps widely spaced is to leave areas of low illuminance between the lamps. This may be seen in Figure 6.20(a). Generally it is found that by spacing the lamps between 1.5 and 2 times their mounting height above the working plane produces an acceptable degree of variation. This is shown in Figure 6.20(b). Using lamps closer together than this increases the installation costs. Imagine installing 452, 60 W lamps to achieve this level of illuminance. Two suitable arrangements of lamps are shown in Figure 6.21.

The arrangement in Figure 6.21(a) uses 48 lamps each with 500 W rating. The total electrical loading $= 48 \times 500 = 24$ kW and the luminous flux provided $= 48 \times 7500 = 360\,000$ lumens.

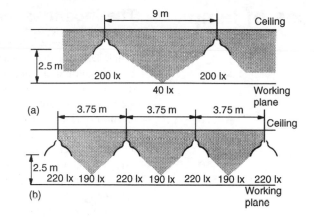

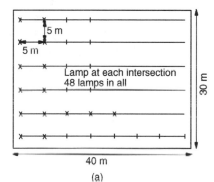

Figure 6.20

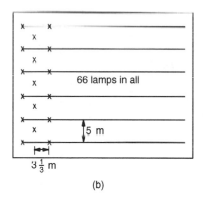

Figure 6.21

An office with a floor area of 10 m × 8 m is to have a uniform level of illuminance at the working surface of 200 lux. Allowing a utilization factor of 0.8 and a depreciation factor of 1.1, determine the number of 1500 mm fluorescent tubes which would be required if each one gives a total output of 2550 lumens when new. Suggest a suitable configuration for the lamps.

The arrangement in Figure 6.21(b) uses 66 lamps each with 300 W rating. The total electrical loading = 66 × 300 = 19.8 kW whilst the luminous flux = 66 × 4140 = 273 200 lumens. This is a closer fit to the requirement than the previous arrangement but requires installing 18 more light fittings at some cost. It is not generally possible to satisfy the requirements exactly without having an asymmetrical arrangement of lamps and possibly having a range of ratings. Such an arrangement gives an unsatisfactory appearance and different-rated lamps give rise to problems when relamping.

Types of lamp

The incandescent lamp

These lamps have a tungsten filament heated within a glass bulb which has either been exhausted to near-perfect vacuum or contains a small amount of inert gas. It radiates energy over a continuous spectrum as shown in Figure 6.2. The proportion radiated as light increases with filament temperature but at best this is only a small fraction of the available energy.

Figure 6.22 shows two possible filament arrangements, one for general service lamps and the other for film projectors. Outputs vary from about 9 lumens per watt for the 40 W size to 14 lumens per watt for the 500 W size. Life and light output vary inversely; increasing temperature and light output reduces life expectancy and vice versa. Thus, running a 240 V bulb at, say, 220 V causes less current to be drawn, the lamp runs cooler, the light output is a lot less but lamp life can be expected to be long.

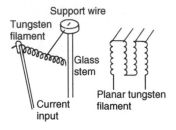

Figure 6.22

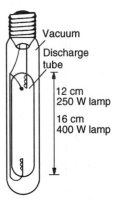

Figure 6.23

The high-pressure mercury discharge lamp

In this lamp the current is caused to flow through a small tube in which there is a little inert gas and a few droplets of metallic mercury. The current flows at first in the gas producing enough heat to vaporize the mercury. The ionization of the mercury vapour produces light. The pressure when hot is about that of the atmosphere, i.e. 1 bar.

The space between the inner discharge tube and the outer bulb is partially evacuated to minimize the heat loss from the lamp (like a vacuum flask to keep liquids hot). This is necessary to keep the mercury in vapour form.

The colour of the light is blue/green with no additional radiation corresponding to red and yellow present. However, there is ultra-violet present and if the inside of the outer bulb is coated with a suitable phosphor, the ultra-violet emission causes this to emit a red component so making the light emitted in total more nearly equivalent to daylight. The uncompensated spectrum is shown in Figure 6.27. The light output is about 40 lumens per watt.

The low-pressure mercury (fluorescent) tube

This lamp is widely used for general illumination. It also contains mercury but at a much lower pressure than the type just described. The discharge tube is also larger physically. A 65 W straight tube is 1500 mm in length whilst the 18 W straight tube is 590 mm in length. Ratings down to about 6 W are obtainable. In addition there are a number of lamps comprising small diameter folded tubes of the form shown in Figure 6.24, with ratings from about 9 W up to 24 W which can be used as direct replacements for incandescent lamps without modifying the light fitting. These have a light output equivalent to an incandescent lamp of about four times their rating. In all these lamps the discharge from the mercury at low pressure

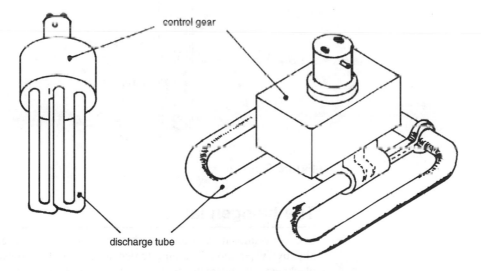

control gear

discharge tube

Figure 6.24

contains a substantial amount of ultra-violet. The inside of the tube is coated with phosphors which emit visible radiations (light) as a result of being excited by the ultra-violet radiations from the gas. The colour can be altered by the selection of phosphors, a technique used in the manufacture of colour television tubes. The light output of the tube varies between 40 and 60 lumens per watt according to colour, 'blue' (north light) tubes giving less light than some of the warmer colours for the same rating. Tubes carry two numbers, one giving their rating, the other denoting the colour of the radiation. They are all temperature sensitive and do not give their full rated output until they have reached their normal working temperature of about 25°C. If they are to be used in situations of permanently low ambient temperature they need a transparent protective envelope to retain their internally generated heat.

The low-pressure sodium discharge lamp

This lamp consists of a U-tube containing a little sodium metal and neon gas at very low pressure. When the supply is switched on, the neon gas conducts some current and red light is emitted. Slowly, as the tube heats up the sodium boils off into the near vacuum of the tube and the ionized sodium vapour emits bright yellow light. The spectrum is shown in Figure 6.27. As with the high-pressure mercury lamp, heat must be conserved and therefore the U-tube is supported within a vacuum flask. If this fractures, the light reverts to the red colour as the sodium solidifies. The light output is about 130 lumens per watt. This lamp is used for road lighting where the colour of the light is not too significant.

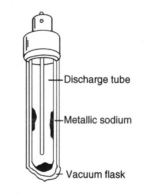

Figure 6.25

The halogen lamp

The conventional tungsten filament decreases in cross-sectional area by evaporation. The tungsten condenses on the bulb wall, so blackening it and reducing light output. The glass bulb is made large and generally burns with its cap at the top. Convection currents within the lamp carry the gaseous tungsten up into the neck of the lamp where it condenses out. Thus the loss of light output due to blackening of the glass is limited.

The halogens are chlorine, bromine, iodine and fluorine. When a halogen is added to the gas inside the lamp and if certain temperature conditions are satisfied, a regenerative chemical reaction may be set up. The tungsten evaporates from the filament as before but, as it approaches the bulb wall and its temperature falls some-what, the tungsten combines with the halide to form a compound, tungsten halide. This compound does not condense on the glass but circulates within the bulb until it comes into contact with the hot lamp filament once more, when the halide is released and the tungsten deposited on the filament. Iodine is the halide currently used, which requires that the filament runs at about 1800°C and the bulb walls be maintained at 250°C. To ensure that the bulb walls attain this temperature, the lamp must be made very much smaller than the conventional type, but this is no longer signifi-cant because there is no blackening of the glass. Because of their reduced size, these lamps can be placed at the centre of focus of a parabolic reflector and they are found in vehicle headlamps and floodlights for public areas. They are commonly used in security systems when they are turned on by heat- and movement-sensitive detectors. Because of the high operating temperature of the bulb wall it is imperative that the lamp is not touched during instal-lation. The grease and dirt deposit in the fingerprint burns on to the glass and impairs light transmission which causes the bulb to overheat at the point touched, leading to premature failure. The light output is between 15 and 25 lumens per watt according to rating (Figure 6.26).

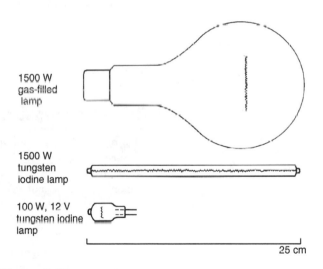

1500 W gas-filled lamp

1500 W tungsten iodine lamp

100 W, 12 V tungsten iodine lamp

25 cm

Figure 6.26

The mercury-metal halide lamp

This is similar to the high-pressure mercury discharge lamp except that small quantities of metallic halides are included in the en-velope. Examples of these are sodium halide and thallium halide. When the mercury discharge attains a sufficiently high tempera-ture, the metals are liberated from their compounds in the same way as in the tungsten halide lamp. In this case, however, the metals are not deposited, but stay in a free form in the mercury discharge. When an electrical discharge takes place in any element the colour produced is unique, as we have already seen in the case of the sodium lamp. The metals included therefore determine the colour of the discharge. The light output from the lamp is that due to that from the mercury discharge in addition to those from the additional metals and the combination is made to give a total output very close to that of sunlight. This lamp is used extensively to light sports stadia. The light output is 70 lumens per watt. As an example, a large stadium from which colour television transmissions are to take place could achieve daylight-equivalent results using 240 lamps each of 1600 W which would give a level of surface illumination close to 1400 lux. The lamp construction is similar to that of the high-pressure mercury type except that there is no auxiliary electrode; starting is by high-voltage impulse.

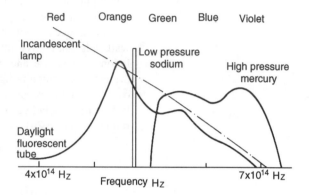

Figure 6.27

The high-pressure sodium lamp

If the pressure in the sodium lamp is increased and the temperature raised to about 1000°C, the discharge becomes much whiter and blends well with light from normal incandescent lamps. This development could not take place until a new material for the tube had been developed. The low-pressure lamp uses a glass tube while this lamp uses an aluminium oxide (Alumina) tube which is capable of withstanding the very high temperature. It is used extensively for floodlighting where the warm, slightly pink colour is very attractive. The light output is about 100 lumens per watt.

Types of luminaire

A selection of luminaries is shown in Figure 6.28.

Types (a) to (d) are used indoors, types (e) to (i) outdoors. They can be diffusing or concentrating and are chosen for these functions, at the same time considering their appearance and the environment in which they are used. Outside luminaires must be weatherproof and as far as possible, vandal proof. The transparent parts are made of pre-stressed glass or impact-resistant plastic. To minimize maintenance, the surfaces should be self-cleaning by the action of wind and rain. Therefore patterns or prisms are on the inner surface of the glass. Electrical connections to them are generally by means of conduit or mineral insulated cables.

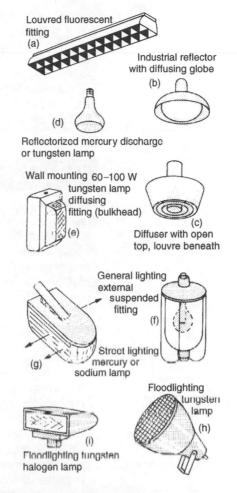

Louvred fluorescent fitting
(a)

Industrial reflector with diffusing globe
(b)

(d)

Reflectorized mercury discharge or tungsten lamp

Wall mounting 60–100 W tungsten lamp diffusing fitting (bulkhead)
(e)

(c)

Diffuser with open top, louvre beneath

General lighting external suspended fitting
(f)

Street lighting mercury or sodium lamp
(g)

Floodlighting tungsten lamp
(h)

(i)

Floodlighting tungsten halogen lamp

Figure 6.28

Activity 6.3

We have seen that electrical discharges in gases produce light much more efficiently than by using that current to heat a tungsten wire. Gas discharge lamps last much longer than incandescent lamps, typically 5000 hours but initially have a much higher cost. Evaluate the economics of using the two types of lamp to produce a set number of lumens over a period of, say, 1 year in a situation of your choice, e.g. in your living room at home, or in an office or factory. Take into account the following factors: (a) cost of discharge lamps and control gear, (b) energy used and the total cost of that energy during their lives, (c) cost of incandescent lamps to give the same amount of light over the period (an incandescent lamp may only last 1000 hours so that you are going to have to buy a lot more of them) and (d) energy used and cost of the energy over the period. With many gas discharge lamps the control gear is separate from the lamp so that when a lamp fails, only the tube needs to be replaced. With others in use control gear and tube cannot be separated. How does this affect your assessment of overall choice and cost? Discuss the implications of construction with regard to recycling and disposal of used equipment. What other factors need to be considered when making the choice?

Activity 6.4

One of the effects of the use of discharge lighting is to reduce the energy requirement of a factory or neighbourhood. Discuss the economics and desirability (or not) of encouraging the use of energy efficient equipment and in particular, lighting as an alternative to the building of power stations and sub-stations to supply increased power as the demand increases. The background to this Activity is that in the USA there have been instances where before getting finance and permission to build a particular installation the supply authority was required to demonstrate that an equivalent electrical capacity could not be freed elsewhere by this means. They have in fact given away, or supplied at very low cost, energy-efficient lighting to demonstrate the point, the result being that the new equipment was found to be not required. The cost of giving away the equipment was more than balanced by the savings in building costs. Research the cost of building power stations and the long-term availability of energy (see also Chapter 1) and use this data together with that evaluated in Activity 6.3 to help formulate a balanced argument.

Daylighting

In most buildings use is made of daylight when available, supplemented by artificial light. The correct balance between the two is difficult especially when the rooms are deep, when the variation in daylighting from a position near a window to that several metres from a window may be considerable.

The amount of daylighting reaching a point inside a building through a window directly from the sky will depend on the area of the sky visible from that point, on the luminance of the sky and the angle at which it arrives at the surface to be lit. As already stated:

$$E = \frac{I}{d^2} \cos \theta° \text{ lux}$$

Figure 6.29 shows how the arc of sky visible is dependent on the distance inside the room. In Figure 6.29(a) at a point P there is 18° of sky visible whilst at point Q there is only 11°. Note how the wall some distance from the building restricts the light input to point Q. If this wall were not there the arc would extend downwards to the first obstruction which might be the window sill or the horizon. In Figure 6.29(b) the effect of a taller window can be seen. More arc of sky is visible from both points. Note also in both cases the difference between the angle of incident light between points P and Q.

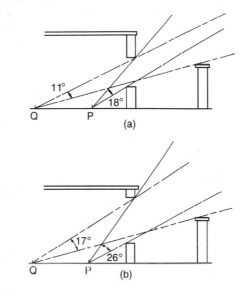

Figure 6.29

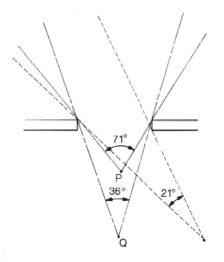

Figure 6.30

Figure 6.30 shows a plan view of the room. Note how the arc of sky visible is limited in the horizontal plane also. At points off the centre line the arc is further reduced.

The luminance of the sky depends on the season of the year, the hour of the day and the distance of the room from the equator, i.e. on its latitude, since the height of the sun above the horizon depends on these factors. The average level of illumination on a horizontal surface in the open, which is a measure of the luminance of the sky, measured at Greenwich using north light only, varies from about 450 lux at 9 a.m. to 1700 lux at midday in January compared with 4150 lux and 5700 lux at the same times in June. In addition, the luminance of the sky will depend on whether it is overcast or not. This may be due to clouds or pollution.

One further factor affecting the amount of light entering a room is the direction in which it is facing, north-facing rooms receiving no direct sunlight. Finally there is a component of light derived by reflection from external surfaces and this depends on the colour and state of these surfaces. This is illustrated in Figure 6.31.

Once light has entered a room by whatever means the final distribution will depend on the colour and state of the decorations, diffused reflection taking place on most surfaces. Figure 6.32 shows a typical illuminance curve for a room lit by north light.

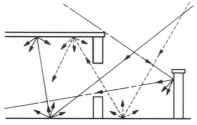

— Light entering the room directly from the sky

-—- Light enering the room by reflection from an external vertical surface

--- Light entering the room by reflection from an external horizontal surface

Figure 6.31

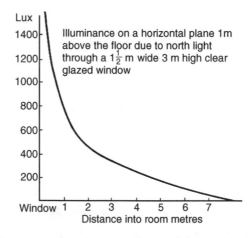

Illuminance on a horizontal plane 1m above the floor due to north light through a $1\frac{1}{2}$ m wide 3 m high clear glazed window

Figure 6.32

Activity 6.5

Investigate the light distribution in a room due to daylighting. Determine the amount of artificial illumination which would need to be added over the inner half of the room to more nearly make the level of illumination uniform across the room.

Use a light meter in the horizontal plane at the height of the working surface to measure the illuminance curve at distances from the window moving back along the centre line of the window directly into the room. Next, at two or possibly three set distances from the wall in which the window or windows are set, move across the room measuring the variations. Note the limiting factors on the light entering the room: external walls, other buildings, trees etc. What are the dimensions of the window(s), how near to the ceiling and floor do they extend?

Comment on your results. What are the problems with very wide variations in illuminance curve within a room? What are the problems with direct sunlight entering a room? Is there a case for having rooms without external windows? What effect would this have on heating requirement? Calculate the required extra illumination which would be required to bring the illuminance of the working surface close to the wall furthest from the window up to the illuminance value at the centre of the room. Suggest ways in which this could be achieved. What are the difficulties of ensuring even partially level illuminance curve over the room under all conditions?

Problems on illumination

1   (a) A metal box has a frosted glass window 1.5 cm^2 in area which is lit from within by a lamp such that it has a luminance of 750 cd/cm^2. What is its luminous intensity in candelas?

(b) (i) If the window is to have a luminous intensity of 900 cd to what value must its luminance be changed? (ii) Assuming that the luminance cannot be changed, suggest a method whereby the required luminous intensity of 900 cd could still be achieved.

[(a) 1125 cd, (b) (i) 600 cd/cm^2 (ii) change area to 1.2 cm^2]

2   The light source in Figure 6.33 has a uniform luminous intensity of 64 cd. The point P on the surface to be illuminated is 4 m from the source. Calculate the illuminance at point P when the surface is in position (a) and then in position (b).

[(a) 4 lux, (b) 3.46 lux]

3   Calculate the illuminance on a horizontal surface due to a single lamp with uniform luminous intensity of 2500 cd mounted 10 m above the surface at points:
(a) immediately below the lamp, (b) 5 m from this point on the horizontal surface and (c) 15 m from the original point on the horizontal surface.

[(a) 25 lx, (b) 17.9 lx, (c) 4.27 lx]

4   Why are incandescent light bulbs often made with an obscured (pearl or opal) glass envelope?

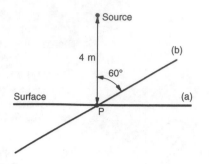

Figure 6.33

5 Why is the tungsten planar filament often used in cinema and slide projectors?

6 Why does it take approximately 10 minutes for the high-pressure mercury and low-pressure sodium lamps to reach full brilliance?

7 What would happen to a working low-pressure sodium lamp if the outer vacuum envelope were removed?

8 What development made the high-pressure sodium lamp possible?

9 What advantages has the tungsten halide lamp in the field of public lighting and automobile engineering?

10 What advantages does the mercury-metal halide lamp have over other gas discharge lamps?

11 What special features should a luminaire have if it is to be used outdoors?

12 A single lamp in a reflector has luminous intensities in the vertical plane as follows:

Angle from the vertical Degrees	Luminous intensity Candela
0	600
20	640
37	800
45	940
53	935
60	600
80	200

The lamp is mounted 7.5 m above a horizontal flat plane. Calculate the illuminance on the plane (a) immediately beneath the lamp, (b) 10 m from that point on the flat plane.

[(a) 10.7 lx, (b) 3.6 lx]

13 A straight row of lamp standards are spaced 30 m apart and each in 8 m high. Each standard carries a lantern which has a polar diagram as shown in Figure 6.34.

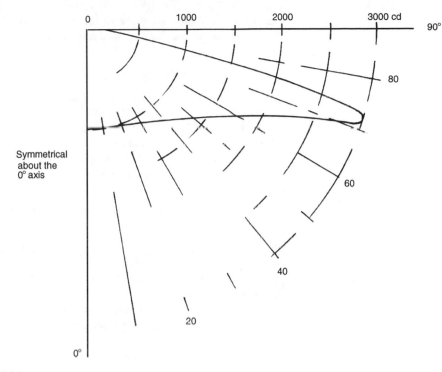

Figure 6.34

Calculate the illuminance 5 m from the base of one standard on a line joining the bases of two standards. Consider the illuminance from the three lanterns nearest to the point.
[$E = 1325/1289 \times 0.223 + 1125/89 \times 0.85 + 3000/689 \times 0.304 = 12.3$ lux. Slight differences permissible due to scaling.]

14   A roadway is illuminated by means of lanterns which have luminous intensities in the vertical plane as given in the table. The standards are 5 m high and 40 m apart.

Angle from the vertical Degrees	Luminous intensity Candela
0	300
10	310
20	330
30	360
40	400
50	450
60	530
65	590
70	685
75	810
80	660
85	0

Calculate the illuminance due to the nearest two luminaires at a point on a line joining two standards, 10 m from one of them.

[$1.97 + 0.106 = 2.076$ lx]

15   A room 20 m × 40 m is equipped with 50 fluorescent tubes each giving a total output of 3400 lumens when new. The coefficient of utilization for the room is 0.56 and a depreciation factor of 1.25 has been allowed. Calculate the level of surface illuminance (a) when all new tubes have just been installed and (b) just before these lamps are due for complete replacement.

[(a) New; light flux $= 95\,200$ lumens, $E = 119$ lux. (b) At end $E = 95.2$ lux]

16   An indoor floor area of 30 m × 50 m is to be lighted to an illuminance of 100 lux. The building has an overall utilization factor of 0.6. A depreciation factor of 1.1 is to be allowed. Consider two schemes:
(a) using high-pressure sodium lamps each of which has a light output of 12 000 lumens when run in for a power input of 130 W.
(b) using 500 W incandescent lamps with an efficiency of 14 lumens per watt.
For each case:
(i) calculate the required number of lamps and suggest a pattern of lamps to suit,
(ii) calculate the cost of energy per annum if the lamps are switched on for 3000 hours and the electricity tariff is £20/kW of maximum demand +4p/kWh of energy used.

[(a) (i) 22.9 lamps (use 24), 4 rows of 6, MD 3.12 kW; (ii) cost £62.4, Energy £374.4.
(b) (i) 39.28 lamps (use 40), 5 rows of 8, MD 20 kW; (ii) cost £400, Energy £2800.]

17   List the ways in which daylight reaches a table top in a room with its only window in the vertical plane.

18   Upon what factors does the illuminance of a surface in a room depend when lit by daylight only?

19   When installing lighting schemes it is often necessary to consider the luminance of the light sources as well as the total lumen requirement. Why is this?

20   In summer time an office 25 m long and 16 m wide receives 100 000 lumens by daylighting. This may be considered to be fairly uniformly distributed over one quarter of the width of the office nearest to the windows.
Calculate: (a) the average illuminance over the area covered by daylighting and (b) the number of fluorescent tubes each giving 3000 lumens necessary to give one half of the illumination in (a) over the *remainder* of the office assuming that daylighting here is negligible. The utilization factor is 0.7.

[(a) 1000 lux, (b) 72 tubes]

Multiple choice questions

1 Luminous intensity is defined as:

   A   light flux per unit area
   B   luminance × area of the source
   C   illuminance × area of a surface
   D   total lux per unit area.

2 A light source has a uniform luminous intensity of 400 cd. It is enclosed in a luminaire which transmits 95 per cent of the light from the source on to a working area. The light flux received by the working area is (in lumens):

   A   5291
   B   4775
   C   380
   D   30.2.

3 A little iodine is added to the gas in a tungsten filament lamp to create the tungsten-iodine lamp. The action of the iodine is to:

   A   make the lamp run cooler
   B   make the lamp run hotter and so increase efficiency
   C   combine with the tungsten evaporating from the filament
   D   decrease the luminance of the source so that it is less uncomfortable to view.

4 Depreciation or maintenance factors associated with the lumen method of lighting calculations depend on TWO of the following. Which are they?

   A   the height of the ceiling
   B   the colour of the decoration
   C   the amount of dirt which collects on walls and lighting fittings
   D   the size of the room
   E   where tungsten lamps are used, the amount of tungsten evaporating from the filament
   F   the shape of the room
   G   the distance between the lighting fittings.

5 Light falls on a surface at 20° to the normal from a source of 600 cd, 3 m distant. The illuminance of the surface is (in lux):

   A   71
   B   563.8
   C   62.65
   D   66.67.

6 A uniform light source causes an illuminance of 450 lux on a surface 15 m away when light rays fall normally on that surface. The luminous intensity of the source is:

   A   2 cd
   B   101 250 cd
   C   30 cd
   D   6750 cd.

7 Illuminance is:

   A   a measure of brightness of a light source
   B   the light being produced by the platinum standard

C the amount of light flux falling on 1 m^2 of a surface

D the number of candelas per square centimetre.

8 A diffusing luminaire is used:

A where a soft lighting effect is required

B where it is necessary to ensure that the light from a fitting is sent in a particular direction

C where the fitting is to be mounted at a considerable height above the working plane

D for road lighting.

9 Ten fluorescent fittings light an area of 10 m × 20 m. Each fitting gives 3500 lumens. The utilization factor is 0.6. The illuminance at the working surface is:

A 175 lux

B 105 lux

C 291.6 lux.

10 At the end of the design life of a lighting installation the total light flux at the working surface is 563 000 lumens. If the maintenance factor is 0.55 and the utilization factor is 0.6, how many lumens at the working surface will there be when a new set of lamps is installed and the illumination is brought back to 'as-new' condition?

A 1 023 636 lumens

B 309 650 lumens

C 938 333 lumens

D 516 083 lumens.

7 Fuse protection

Summary

This chapter explains the action of circuit protection using fuses. It examines the construction of fuses and explains how the energy input to a circuit under fault conditions is limited. Arcing time and pre-arcing time are defined and a typical current-time curve for an HRC fuse is examined.

Fault level

When considering circuit protection, the fault level at the point to be protected must be taken into account. The fault level is the number of volt-amperes which would flow into a short-circuit if one occurred at the particular point. On a 415 V three-phase street main this could be around 2000 kVA, which represents a current of nearly 3000 A. The fault level is limited by the impedance of all the apparatus between the point of the fault and the supply generator. When such a current flows a protective device such as a fuse should operate to cut off the supply. The value of current which would flow if nothing was done to limit it is called the 'prospective current'.

At a similar voltage within a factory where the distance from the supply transformer may be less and the transformer itself larger, the fault level could rise to 18 000 kVA (18 MVA) which represents a current of 25 000 A. The nearer the fault to the supply generators, the larger the fault level. A fault on the 400 kV system can give fault levels in excess of 30 million kVA (30 000 MVA), which is effectively a very large part of the total generating capacity of the UK.

Earthing

Both the IEE Regulations and Electricity Regulations require one point on a transformer-fed system to be earthed. Such earthing provides a return path for earth current, so facilitating the clearance of faulty circuits.

Power transformer secondaries are star-connected and are earthed in exactly the same manner as the generators (Figure 7.1).

When a fault occurs on a piece of equipment such that the live conductors or part of the winding of the equipment comes

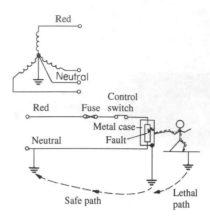

Figure 7.1

into contact with the metal casing of the equipment there exists a potentially dangerous situation. If the casing is not connected to earth a person standing on the ground touching the casing will receive an electric shock as the potential difference between that of the casing and earth causes a current to flow through the body. It only requires a few milliamperes along a route including the heart to cause death. Earthing the equipment allows current to flow through the earth connection back to the supply transformer neutral or star point. If the impedance of the earth loop is small, sufficient current will flow to melt the fuse in the supply line.

A high earth loop impedance is dangerous since the metal case will remain connected to the supply if insufficient current flows to cause its clearance.

In certain situations equipment is run without an earth wire. Such equipment is completely enclosed in a layer of insulating material or so disposed in an insulated housing that it is impossible for the housing to become alive or for a person to push a finger in and so touch live metal. A television set in a plastic or wooden cabinet is an example of this type of construction.

Fuse protection

Two types of fuse are in general use:

1   the semi-enclosed, rewireable fuse
2   the high rupturing capacity fuse, or cartridge fuse (HRC).

Marked on each cartridge fuse and quoted for each size of fuse wire is a current rating. The current rating is that value of current which the fuse can carry indefinitely without melting or deteriorating.

The rated minimum fusing current is the least value of current which will actually cause the fuse element to melt.

$$\text{Fusing factor} = \frac{\text{Rated minimum fusing current}}{\text{Current rating}}$$

This factor may lie between approximately 1.2 and 2.0. With a fusing factor of 1.5, a fuse rated at 10 A will require a current of $1.5 \times 10 = 15$ A to operate.

Operation of fuses

When current flows in a fuse element, heat is produced. If that current is less than the current rating of the fuse then the body of the fuse is able to dissipate the heat to the surroundings, usually into the air. When the rate of producing heat is equal to the rate of dissipating it, then the temperature remains constant at a value lower than that necessary to melt the fuse element.

When a current slightly greater than the minimum fusing current flows, heat is produced in the element faster than it can be dissipated; after a period of time, the element melts and the circuit is interrupted. In the event of a short-circuit or possibly an earth fault, the prospective current is extremely large and heat is generated in the element at such a high rate that there will be no time for the body of the fuse to be heated and heat to be dissipated and the element melts extremely rapidly, circuit clearance times being extremely short. When operating in this manner the fuse is said to have 'inverse time characteristics'. This means that the larger the current, the shorter the time taken to clear the circuit.

The semi-enclosed rewireable fuse

This comprises a ceramic fuse base with an asbestos or similar liner into which the fuse carrier is slotted. The fuse has male contacts of brass and pushes into spring-loaded female contacts of brass or berylium copper in the base. The fuse element is tinned-copper wire, the diameters for particular currents being specified in the IEE Regulations. These fuses are the subject of BS 3036. (See Figure 7.2.)

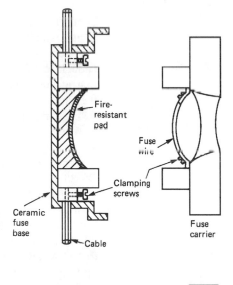

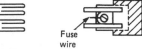

Contact arrangement

Semi-enclosed rewireable fuse

Figure 7.2

The fuses are manufactured with ratings up to about 100 A and are suitable for use in alternating-current circuits operating at up to 250 V. In clearing, the fuse element melts and so interrupts the circuit current. The copper of the element partly vaporizes and partly spreads itself over the surface of the carrier. It operates perfectly satisfactorily with overloads and fairly small fault currents. However, with very high prospective currents this melting takes place explosively and current may continue to flow in the metallic vapour present. The danger here is that the fault would have to be cleared by a much larger protective device further back in the network, possibly at the main circuit breaker. Before this can operate the fuse carrier and possibly the fuse board may be suffering severe damage. As the main device operates, other consumers will lose their supply. Rewireable fuses are also open to abuse in that they can be rewired using almost any conductor, such as a nail or hairpin. The temptation is to upgrade the size of fusewire in a circuit which suffers consistent overcurrents rather than investigate the reason. The IEE Regulations recommend that wherever possible a cartridge-type (HRC) fuse be used. At the fault level found in domestic situations, a semi-enclosed rewireable fuse is generally quite adequate.

The high rupturing capacity (HRC) fuse

The HRC fuse consists of a cylindrical ceramic body to the ends of which are attached brass or copper end-caps. The fusible element is generally of silver, since this is the best known electrical conductor.

Three forms of the element are shown in Figure 7.3. On one of them a small bead of soft solder has been deposited. The body is tightly packed with very fine quartz grains which keep the element in place at all times, including whilst melting, and gives controlled heat conduction away from the element. With currents just large enough to cause melting, the element with the solder bead melts first at this point, whilst the other type melts at one or more of the narrow sections. The presence of the solder bead gives a lower fusing factor than is possible by most, if not all, other methods.

Where very large currents are involved, the heat produced explosively melts the silver which chemically combines with the filling to form a compound, whose resistance increases rapidly as it cools. BS 88 covers fuses for working up to 1000 V AC with ratings of several hundred amperes. They may be used to protect circuits with extremely high prospective currents, 80 kA being possible. Tests have been carried out on some types, clearing currents of 100 kA. Cartridge fuses for domestic premises, generally only up to 100 A rating, are covered by BS 1361 and prospective currents up to 33 kA can be dealt with.

The small cartridge fuses found in domestic ring-main plugs have ratings between 3 A and 13 A and can interrupt currents up to 6 kA. They comprise a ceramic tube, quartz-filled with a single wire element, and employ the solder bead principle to give rapid clearance at low values of overcurrent.

Small cartridge fuse links for telecommunication and light electrical apparatus are rated from about 50 mA to 5 A at 250 V. They are covered by BS 2950. Other miniature fuses appear in BS 4265.

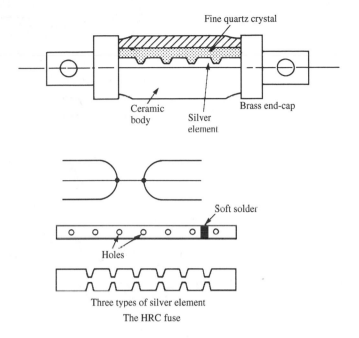

Three types of silver element
The HRC fuse

Figure 7.3

Miniature fuses vary in speed from 'super quick-acting' to 'super time lag'. The quick-acting types have a single strand wire element of silver, copper or nickel-chromium often with a small part of its length having reduced cross-sectional area. The time lag varieties allow a surge to pass without rupturing and so are termed 'anti-surge' types. Two of these are shown in Figure 7.4. One of them has a single wire with a soft solder bead situated about halfway along. With small prolonged overcurrents, the wire melts at the solder bead. With large overcurrents, the whole wire vaporizes. In the other, with spring-assisted break, at low overcurrents the soldered joint melts, allowing the spring to pull back to give a good wide break. With severe overcurrents the straight section of the element (called the 'heater wire') vaporizes and, again, the spring rapidly increases the gap. These types are not quartz-filled.

Activity 7.1

Two of the British Standards relevant to fuse manufacture and operation are BS 88 and BS 3036. Obtain copies of these and see how the two types are different. Report on the methods of construction. Note the physical sizes of various ratings of fuse. Why would a 30 A HRC fuse for use on a main distribution board be significantly larger in size than a similar rated fuse for use in a domestic ring-main circuit? From a supplier's catalogue or otherwise get prices for various sizes and observe the steep increase in price as the rating is increased. Relate this to the use of

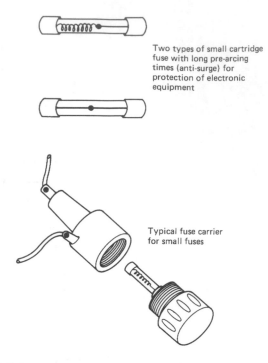

Two types of small cartridge fuse with long pre-arcing times (anti-surge) for protection of electronic equipment

Typical fuse carrier for small fuses

Figure 7.4

miniature circuit breakers for domestic use. What are the advantages of the use of MCBs? Investigate the term 'fusing factor'. Why is it not possible to manufacture any type of fuse without such a factor? When using fuses to BS 3036 an additional de-rating factor has to be used as well as those for ambient temperature and for grouping. What is this factor? Why has it to be used? (See also in Chapter 5, Permitted volt drops in cables.)

Activity 7.2

Carry out a load test on a 5 A rewireable fuse. Use a rheostat to load the low-voltage output from a variable transformer (Variac). Include a rewireable fuse complete with its base and carrier and an ammeter in the secondary circuit. By gentle variations in variac and rheostat settings it will be possible to establish the rated current of the fuse wire in the fuse carrier. If a contact thermometer is available measure the temperature of an inner surface of the carrier as close to the fusible element as possible. Otherwise, after some minutes de-energize the circuit, remove the carrier and experiment by feeling to see how warm it has become. Increase the current in, say, 0.5 A increments, waiting some time before testing the temperature and making a further increase. Record the value of current at which the element melts. With a new fuse wire installed, starting from cold, immediately establish this current and measure the time taken for it to melt. Start again with a new wire, and measure how long it takes to melt with a further increase in current, say 25 or 50 per cent more.

For comparison between the types the tests can be carried out using an HRC fuse, possibly a 5 A cartridge in a 13 A plug top.

Comment on the fuse performance. How repeatable are the values of current and time for the rewireable fuse? What advantages are there in using the cartridge type? What external factors influence the performance of a rewireable fuse? Discuss the possible repercussions of a failure of a rewireable fuse to clear a short-circuit fault in a fuse board of your choice: in your home, at your place of work etc.

Fuse characteristics

When a much larger current than normal full load is allowed to flow in a circuit, the cable may suffer damage in two ways. Firstly, the excess current will generate a large amount of heat. Secondly, currents flowing in adjacent conductors set up mechanical forces (Chapter 5, Circuit damage due to overcurrents). These forces may be large enough to disrupt the cable.

For these reasons, the fuse must be capable of interrupting large prospective currents before they reach their first maximum value. The fuse must begin to melt during the first quarter-cycle of current and so prevent the full prospective value being realized. Figure 7.5 shows the prospective current in a circuit as a dotted line. This is the current which would flow in the circuit if the fuse were not present and nothing was done to limit the fault current.

After a short time the energy dissipated in the fuse element by the large and rapidly increasing current is sufficient to start it melting. The time taken up to this instant is called the 'pre-arcing time'. Once the element begins to melt, the resistance of the fuse rapidly increases and the circuit current decreases. During this time there is an arc within the fuse and this period is called the 'arcing time'. The pre-arcing time plus the arcing time is the total clearance time.

Each manufacturer makes a very large range of fuses, each with a different characteristic to suit a particular requirement. It is only possible to deal with these in general terms. To determine the precise time for clearance of a particular fault, the actual fuse manufacturer's data must be consulted. Figure 7.6 shows a general set of inverse time characteristics for a fuse.

Looking at the curve for the 20 A fuse, for example, it may be seen that it will carry almost 30 A for ever without clearing. This tells us that its fusing factor is in the region of $30/20 = 1.5$. As faults with increasing prospective currents are thrown on to the fuse, the pre-arcing times decrease. Following the lines marked (a) in the figure we see that, with a prospective fault current of 80 A, the fuse has a pre-arcing time of 1 second. This is after 50

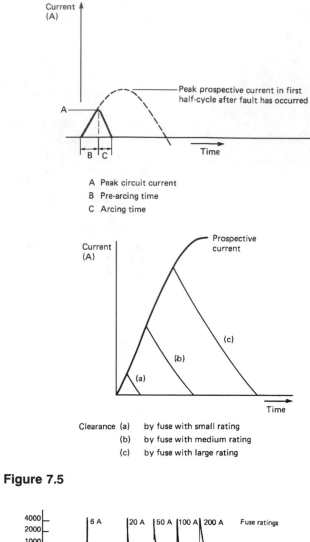

A Peak circuit current
B Pre-arcing time
C Arcing time

Clearance (a) by fuse with small rating
        (b) by fuse with medium rating
        (c) by fuse with large rating

Figure 7.5

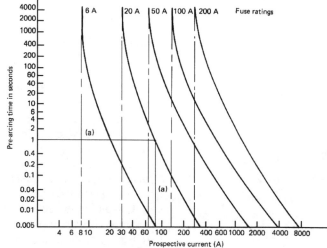

Figure 7.6

complete cycles of the current. Proceeding down the curve to its intersection with the bottom axis it is seen that with a prospective current slightly in excess of 300 A the pre-arcing time is 0.005 seconds, which is one quarter of a cycle. For prospective currents greater than 300 A the pre-arcing times will be even shorter, total clearance being achieved in less than one half cycle.

We have been considering the fuse as a device which limits the size of a potentially large current or limits the time during which a current only marginally greater than the normal full-load value is allowed to flow. Another way of viewing the fuse's action is to consider the amount of energy being allowed into the protected circuit. If we consider a 240 V supply feeding a load through a cable, the total resistance of the circuit (including that of the fuse) being 0.05 Ω, in the event of a perfect short-circuit at the end of this cable the current could theoretically rise to a value $240/0.05 = 4800$ A. If this current flows for a complete cycle the energy input to the circuit can be calculated as follows:

Power input $= I^2R = 4800^2 \times 0.05 = 1.152 \times 10^6$ watts
Energy $=$ power $\times$ time and time for one cycle $= 0.02$ s
(at 50 Hz)
Energy $= 1.152 \times 10^6 \times 0.02 = 23\,040$ joules.

Now suppose that the resistance of the fuse itself is 0.01 Ω and that of the circuit is 0.04 Ω. Consider the quantity I^2t.

$I^2t = 4800^2 \times 0.02 = 460\,800$ A^{2}s. Since this quantity does not contain the resistances of the two parts of the circuit, it applies both to the fuse and to the cable.

Energy used in clearing the fuse $= I^2Rt = (I^2t) \times R$

$$= 460\,800 \times 0.01$$

$$= 4608 \text{ joules}$$

Notice that this is the actual energy used in melting the fuse element.

Energy supplied to the cable $= I^2Rt = (I^2t) \times R$

$$= 460\,800 \times 0.04$$

$$= 18\,432 \text{ joules.}$$

Total energy supplied $= 18\,432 + 4608$

$$= 23\,040 \text{ joules as before.}$$

The fuse manufacturer could tell us the actual energy requirement to melt the fuse, but it is just as useful to quote the I^2t value since this is the same for both fuse and circuit. The value quoted assumes that there is no time for the element to pass heat to the surrounding air, i.e. the value quoted is for short-circuit conditions.

Fuse manufacturers quote values of I^2t input (1) to cause arcing to commence and (2) to clear the fuse totally. Provided that the I^2t requirement to clear the fuse is less than that which would cause damage to the associated circuit, all is well.

This is particularly important in high-power electronics employing devices such as thyristors. These can carry extremely high currents but are physically very small. The amount of energy required to destroy them is quite small and a special range of very fast HRC fuses has been developed to protect such devices.

As an example, consider two 20 A fuses: (1) a high-speed type with I^2t value = 250 A²s and (2) a motor protection fuse with $I^2t = 3000$ A²s. In each case the circuit being protected develops a short-circuit fault with resistance 0.05 Ω.

Fuse (1) will allow $I^2t \times 0.05 = 250 \times 0.05 = 12.5$ joules into the circuit whereas fuse (2) will allow $3000 \times 0.05 = 150$ joules to pass.

Which fuse is selected will depend on knowing the energy level which will damage the circuit. If it is cable, 150 joules is unlikely to harm it. If the fuse is connected to a solid semi-conductor device, it may well be that 150 joules would damage it and the energy input would have to be limited by using fuse (1).

Test your knowledge 7.2

Using the characteristics in Figure 7.6 determine for the 50 A fuse:

(a) the pre-arcing time for a current of value approximately 140 A

(b) the value of current to give a pre-arcing time of 0.01 second.

Problems on fuse protection

1 What is meant by the term 'Fault level' as applied to an electrical circuit?

2 Why has the earth loop impedance of a fuse-protected circuit to be low?

3 Explain why a circuit 'prospective current' would most likely be much higher in a factory with its own supply transformer than at a house in a street where the supply transformer is some distance away.

4 Why is the fusible element of an HRC fuse often made from silver?

5 What is the function of the solder bead added to some fuse elements?

6 A fuse has a rating of 15 A and a fusing factor of 1.3. What is the minimum value of current which will cause the fuse to operate?

[19.5 A]

7 A circuit protected with a 20 A HRC fuse will have a smaller conductor size than a circuit protected by a 20 A semi-enclosed fuse. Why is this?

8 Why is it necessary to interrupt fault currents before they rise to their full prospective value?

9 Sketch a curve showing (a) 'pre-arcing time', (b) 'arcing time' and (c) 'total clearance time' for a fuse. Explain what is happening within the fuse during periods (a) and (b).

10 Sketch the construction of (a) an HRC fuse for a power circuit, (b) a cartridge fuse for a piece of low-rated electronic equipment and (c) a semi-enclosed fuse for use in domestic premises.

11 What does the term 'anti-surge' mean with reference to small cartridge fuses?

12 Why are extremely fast-acting HRC fuses used in power semi-conductor circuits?

13 Why would a fast-acting HRC fuse be unsuitable for the protection of an induction motor? What characteristics would be required here?

14 Select, giving reasons, fuses to protect the following equipment:

(a) a 50 kW induction motor,
(b) the lighting on one floor of a domestic dwelling,
(c) a piece of electronic equipment used on the bench of a laboratory,
(d) a semi-conductor rectifier inside a large motor speed controller,
(e) a domestic kettle.

15 What are 'inverse-time' characteristics. Explain why a fuse which takes many minutes to clear with a small overcurrent can clear within one half-cycle of the alternating input when a short-circuit occurs.

16 A fuse protecting a circuit has a resistance of 0.005 Ω. A fault develops on the circuit which has a resistance of 0.1 Ω. The fuse has an I^2t value of 2500 A²s for total clearance. Determine:

(a) the energy used in melting the fuse element,
(b) the energy entering the faulty circuit.

[(a) 12.5 joules, (b) 250 joules]

17 A fuse manufacturer quotes a value of I^2t of 0.8×10^6 A²s for the total clearance of an HRC fuse. A fault occurs on a circuit being protected by this fuse such that the circuit resistance falls to 0.02 Ω. How much energy will enter the circuit whilst the fuse is clearing?

[16 000 joules]

18 What is the necessary relationship between total clearance I^2t of a fuse and the permitted level of I^2t of the protected equipment, to give satisfactory protection?

Multiple choice questions

1 A fuse has a fusing factor of 1.4 and a rating of 15 A. It will clear when carrying a minimum current of:

A   15 A
B   10.7 A
C   21 A.

2 The term 'inverse time operation' covers operation:

A   when the fuse has time to dissipate heat to its surroundings
B   under short-circuit conditions
C   at currents slightly in excess of those required to cause the element to begin to melt
D   with currents having values between its rating and its fusing factor times its rating.

3   Rewireable fuses have elements made from:

A   copper
B   silver
C   gold
D   nickel-chromium.

The element in a rewireable fuse is plated with one of the following materials:

A   silver
B   gold
C   tin.

4   The HRC fuse cartridge employs an element most often made from:

A   gold
B   copper
C   silver
D   nickel.

The space within an HRC cartridge fuse is packed with:

A   quartz sand
B   magnesium carbonate
C   cross-bonded polyethylene.

5   A semi-conductor circuit is to be protected using a high-speed HRC fuse. The I^2t rating of the semi-conductor is 250 A^{2}s. Fuses are available with total I^2t let-through of

A   250 A^{2}s
B   500 A^{2}s
C   200 A^{2}s.

Select the one most likely to give satisfactory protection against short-circuit faults.

6   In some HRC cartridge fuses the element has a bead of solder at some point along its length. The solder bead:

A   holds together two separate pieces of fuse wire, one piece joined to each end cap
B   provides a weak spot where the element can melt first
C   provides a section with a large heat-storing capability so that the fuse wire melts in two places, one on each side of the bead.

7   The anti-surge fuse is so called because it has the following property:

A   It has a high inductance in the form of a coiled spring at one end so that when a system surge occurs it prevents the rapid rate of rise of current which the surge would cause.
B   It has a long pre-arcing time which allows the surge to pass without affecting the fuse.
C   It has a short pre-arcing time so that the peak value of the surge is limited.

8 Tariffs and power factor correction

Summary

This chapter examines the costs involved in the supply and distribution of electrical energy and the charges made for energy using the basic tariffs. Power factor correction is explained and the savings which can be made are calculated.

Generation and transmission costs

The cost of generation and transmission of electrical energy is divided into two parts:

1 Capital charges
2 Running charges

Capital charges

In order to build a power station or a transmission line money is borrowed and interest paid annually. Borrowing may be from an institution such as a bank when the interest will be determined by the bank and this together with some repayment of capital will be made according to a plan. In the case of a public company the money may be the result of a share issue when the 'interest' paid is in fact a dividend to shareholders, the amount of which will depend upon the profitability of the enterprise. Capital is not normally repaid in this case. To regain money spent on ordinary shares it is necessary to find a buyer. The company borrowing the money will generally make a depreciation allowance which means that the book value of the power station or transmission system is decreased year by year. This amount (more, or less; dependent on the outlook for the future) will be invested in new plant. Interest is payable whether the plant is used or not and dividends are paid based at least in part on the profitability of the company. To sum up, the money borrowed buys new plant which has to generate or transmit electricity. Customers pay for the energy and for transmission and the income gradually pays for new plant (depreciation allowance), pays the lenders their dividends, possibly pays off some fixed loans and runs the operating company.

Running charges

In order to run a power station, staff must be paid, fuel purchased and repairs carried out. On overhead lines and underground cables inspections, tests and repairs must be carried out. These costs are very nearly proportional to the amount of energy sold. The cost of losses in the system are also included in the running charges. The consumer has to pay running charges made up of (1) generation cost, (2) transmission costs and (3) distribution costs.

Background information

Electricity is generated, in the main, in a number of large power stations; energy is transmitted through the grid system to a number of regional companies who buy it in bulk and then sell it on to their customers. The industry which previously was made up of literally hundreds of small undertakings, some linked with the grid, others not, was nationalized in 1948 and after a number of name changes it became the Central Electricity Generating Board. Electricity Boards were created to retail power. The Electricity Act 1989 provided for the privatization of the supply industry and power stations previously owned by the CEGB were transferred to National Power plc, PowerGen plc and Nuclear Electric plc. Nuclear Electric has now been further divided. The transmission system is now operated by The National Grid Company and privatized regional electricity companies sell on the power to most consumers. It is now possible for users with an average demand of 100 kW over a three-month period to establish contracts with a supplier of their choice. In Scotland there are Scottish Power plc, Hydro Electric plc and Scottish Nuclear Electric plc.

The National Grid Company (NGC) is responsible for matching supply and demand and in deciding the charges to be made to the retail companies and paid to the generators. Since electricity cannot be stored, at all times supply must match demand. Each day estimates are made of the requirement for energy during the following day on a half-hourly basis, from the hour to half past and from half past to the next hour. The demand will be different for each half hour as will be the costs of the energy provided during that time. During the night, for example, only the plant with the lowest operating costs need run since the demand is low generally. As the load picks up for breakfast time more plant needs to be started up and this may well have a higher fuel charge. On a cold winter's day late in the afternoon when street lights are on, factories still working and evening meals are being prepared the high (peak) demand may well require that some low efficiency plant be run just for a short period. This may well have a very high fuel charge. The NGC calls on the electricity generators as required and they offer plant capacity quoting their 'bid price'. According to these bids the plant is selected and run and the buying price is computed according to the highest-cost supplier on line in a particular half hour. This is known as the 'pool price'. The values of the half-hourly demands and costs are recorded by generators and the regional electricity companies. The price differences over the day and over a year are considerable and there are also contracts outside the pool with suppliers to try to even out these differences.

Total cost of electricity

Charges to consumers are:

1 All the plcs mentioned above have been funded by share issues and the charges discussed in **Capital charges** above are relevant. These charges have to be met in one way or another and ways of charging various groups of customers will be investigated later.
2 Fuel and running charges. This is the amount discussed above under Pool price.
3 Transmission charges. The NGC makes a charge to transmit power and this covers all the costs of maintaining the grid and where necessary installing new equipment.
4 The supply companies charge to distribute electricity and to offer services to their consumers. They read the meters and bill consumers.

Power factor

Most loads on the electricity supply system comprise resistance and inductance in series so that the supply current lags on the voltage by an angle ϕ' as shown in Figure 8.1.

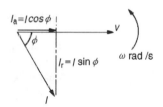

Figure 8.1

The amount of this current which is in phase with the voltage and therefore of doing work is called the active component of current I_a.

Power in the circuit $= VI_a$ watts
The product $V \times I$ is termed volt-amperes.

$$\frac{\text{True power}}{VI} \text{ is the circuit power factor}$$

Hence, power factor $= \dfrac{VI_a}{VI} = \dfrac{I_a}{I}$ which from Figure 8.1 can be seen to be equal to $\cos\phi$.

Also in Figure 8.1, the vertical side of the current triangle I_r can be found since $I_r/I = \sin\phi$ so that $I_r = I\sin\phi$.

In a circuit comprising pure resistance, all the current is in phase with the voltage so that $I = I_a$ and $I_r = 0$. I_r is only present when the circuit has reactance and is therefore known as reactive current.

Taking the current triangle in Figure 8.1, multiplying each side by V, the circuit voltage, gives a similar triangle showing the relationship between power, volt-amperes and volt-amperes reactive as the product VI_r is called.

Since the current lags on the voltage in this case these are known as lagging volt-amperes reactive.

[Notice carefully in Figure 8.2; I is the symbol for current whilst A is the unit of current. We write $I = 5$ A, for example. Similarly $VI_r = 250$ VA$_r$.]

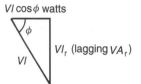

Figure 8.2

Maximum demand

As we have already discussed, payment must be obtained from consumers to cover capital costs of plant which are independent of the number of hours the plant runs or of the electricity generated. Consider a small system of, say, 100 000 kW capacity, the total capital charges on which are £1.5 million per annum. This represents £15 for each kilowatt of plant capacity per annum. Thus 100 000 consumers each with a 1 kW simultaneous demand should each pay £15 per annum to cover the capital charges. Whether the consumer leaves the equipment on for the whole year or just for the one hour, the amount of plant involved is the same and the charge is the same. There would of course be a difference in the energy used and the running charge between the two cases.

When the supply companies purchase energy from the generators the amount of plant involved at any particular time and specifically the maximum amount is determined by the half-hourly recordings as described above. So too the maximum demand of a consumer may be determined. As an example, should the maximum reading over one half hour during a set period of, say, one month be 350 kWh this is equivalent to 700 kWh in one hour or a maximum demand of 700 kW. As an alternative to continuous half-hourly recordings the consumer may have a meter which simply retains the highest reading achieved during a given period.

Measurements of the reactive component involved may also be made when kWh and kVA$_r$h are recorded. (See Figure 8.2, Wh = $VI \cos \phi \times$ time (hours) and VA$_r$h = $VI_r \times$ time (hours).

Diversity factor

Within a factory or premises, not all the installed equipment will be working simultaneously.

Diversity factor =

$$\frac{\text{Demand of equipment actually connected at any instant}}{\text{Maximum demand}}$$

Load factor

Load factor =

$$\frac{\text{Energy consumed in a given period}}{\text{Energy that would have been consumed had the maximum demand been sustained during the period.}}$$

Cost of electrical energy

Domestic and Commercial

Because of the high cost of special metering and the relatively low demand involved the capital charges are recovered from domestic and commercial consumers by using a front-end-loaded tariff with a possible fixed charge in addition. The use of this type of tariff is best illustrated using examples.

Example 8.1

(a) Calculate the cost of electricity per quarter supplied to a small business which has a load factor of 0.05 (5%) and a maximum demand of 10 kW for the following tariff:
Standing charge per quarter (91 days) = £9.50. First 1000 kWh/quarter = 9p/kWh, all other energy at 6p/kWh.

(b) What is the average cost/kWh?

(c) What would the total cost and the average cost per kWh be if consumption increased by a factor 1.5 times?

(a) Load factor $= \dfrac{\text{Energy consumer per quarter}}{\text{Maximum demand} \times \text{hours per quarter}}$

$$0.05 = \frac{\text{Energy consumed}}{10 \text{ kW} \times 91 \text{ days} \times 24 \text{ hours}}$$

Energy consumed $= 0.05 \times 10 \times 91 \times 24 = 1092$ kWh

1000 kWh at 9p/kWh = £ 90.00
92 kWh at 6p/kWh = £ 5.52
Standing charge = £ 9.50
Total cost = **£105.02**

(b) Average cost per kWh = 10 502 pence/1092 kWh = **9.62p/kWh**

(c) With usage increased by a factor 1.5 times, kWh consumed = 1.5 × 1092 = 1638 kWh. Charges for the first 1000 kWh are unchanged.
The next 638 kWh at 6p/kWh = £38.28.

Total cost = £90 + £38.28 + £9.50 = **£137.78**.

Average cost/kWh = 13 778 pence/1638 kWh
= **8.41p/kWh**

Test your knowledge 8.1

A family uses 450 kWh in a winter quarter. The quarterly tariff is: Standing charge £8.50. All energy used 8p/kWh. Determine:

(a) the total charge for the winter quarter

(b) the average cost per kWh.

Test your knowledge 8.2

1 A small bakery business uses 6500 kWh in a quarter year. It is charged on a tariff:

Standing charge £9.50
The first 1000 kWh in each quarter charged at 10p/kWh
All other units at 7.5p/kWh

Calculate the total bill for the quarter. What is the average cost per kWh?

2 As an alternative to the above, a tariff offering a reduced rate for electricity used at night is on offer. This tariff is:

Standing charge £13
Each unit used at night charged at 2.75p/kWh
First 1000 kWh used during the day charged at 10.5p/kWh
All other daytime units charged at 8p/kWh.

Because of the increased losses due to keeping the premises warm and water hot all night the total energy usage increases slightly and the division is 4000 kWh used in the day and 3000 kWh used at night. What is the total cost of electricity on this tariff? Would the change-over be worthwhile?

Industrial

A number of variations on tariffs exist based on maximum demand and some of these are illustrated using further examples.

Example 8.2

A factory has a maximum demand of 200 kW, a load factor of 0.4 (40%) and an operating power factor of 0.7. Calculate the annual cost of energy and the average price per kWh on the following three tariffs:

(a) Maximum demand charge £20/kW. Running charge 3.5p/kWh
(b) Maximum demand charge £17/kVA. Running charge 3.5p/kWh
(c) Basic maximum demand charge £20/kW increased by a factor 0.1 (10%) for each 0.1 that the power factor is worse than 0.9. Running charge 3.5p/kWh.

The MD charges are on an annual basis.

(a) $$0.4 = \frac{\text{Energy consumed}}{\text{Maximum demand} \times \text{hours in the year}}$$

Energy consumed = 0.4 × 200 × 365 × 24 = 700 800 kWh
Maximum demand charge = £20 × 200 = £4000
Energy charge = 700 800 × 3.5p = £24 528

Total cost = £24 528+4000 = **£28 528** (= 2 852 800 pence)

Average cost per kWh = cost/number of kWh used
= 2 852 800/700 800
= **4.07p/kWh**

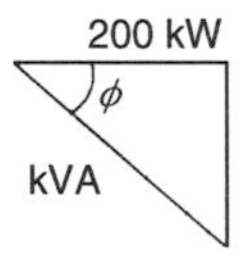

Figure 8.3

(b) Power factor = 0.7 = $\cos\phi$ (see Figure 8.3)

$$\frac{200}{\text{kVA}} = 0.7$$

$$\text{Maximum demand in kVA} = \frac{200}{0.7} = 285.71$$

Maximum demand charge = £17 × 285.71 = £4857.07
Energy costs are as before = £24 528
Total cost = **£29 385.07**
Average cost/kWh = 2 938 507/700 800 = **4.19p/kWh**

(c) The power factor $= 0.7$. This is 0.2 worse than 0.9 so that the maximum demand charge is increased by a factor 0.2.

The maximum demand tariff $= £20 + (0.2 \times £20)$
$$= £24/\text{kW}$$

Maximum demand charge $= £24 \times 200 = £4800$

Energy cost $= £24\,528$

Total cost $= £\textbf{29\,328}$

Average cost/kWh $= 2\,932\,800/700\,800 = \textbf{4.18p/kWh}$

Sometimes instead of just using maximum demand, an industrial consumer is asked to declare what maximum capacity he expects the supply company to be able to provide during a given period of, say, one month, and a charge is made for having this available. Such a charge is known as the *availability charge*. The next example incorporates this together with a charge for kVA$_r$h. This last charge can be avoided by power factor improvement as discussed in further sections.

Example 8.3

The following charges are for the month of January and are for an industrial consumer taking a supply directly from the 33 kV system. (Note that they will be different in the summer when there may well be no maximum demand charge since there will be spare capacity on the system.)

Availability charge $= £1.10$ per kVA

Reactive charge for each kVA$_r$h is excess of one half of the number of kWh supplied $= 0.3$p

Maximum demand charge $-$ £6.00 per kW

Unit charge: (a) for each kWh of the first 100 per kW of maximum demand $= 5$p

(b) all kWh in excess of (i) $= 3.5$p

The factory has declared that it requires an availability up to 500 kVA.

Actual MD during January $= 450$ kW
Energy consumed $= 105\,600$ kWh
Reactive consumption $= 65\,000$ kVA$_r$h

Determine the total electricity charge for January and the average cost per kWh.

Availability charge $= 500 \times £1.10 = £550$

Maximum demand charge $= 450 \times £6 = £2700$

One half of kWh consumed $= 105\,600/2 = 52\,800$

Reactive component charge $= 0.3$p for each kVA$_r$ h in excess of this value

Charge $= (65\,000 - 52\,800)$ kVA$_r$h $\times 0.3$p $= £36.60$

Energy charge at higher rate is for 100 kWh for each of the 450 kW of maximum demand.

Charge $= (450 \times 100) \times 5p = £2250$
Other units $= 105\,600 - 45\,000$
$\qquad = 60\,600$ charged at 3.5p/kWh
Charge $= 60\,600 \times 3.5p = £2121$
Total cost $= £(550 + 2700 + 36.60 + 2250 + 2121)$
$\qquad = £7657.60$
Average cost $= 765\,660p/105\,600$ kWh $= 7.25p/kWh$

Activity 8.1

Find out the tariff on which your place of work pays for its electricity. Try to inspect the metering; is it simply a kWh meter with possibly a maximum demand indicator or is there additional metering for reactive volt-ampere hours? If so, what is its purpose? Is there provision for using power off-peak? What are the implications of this? Why would you not expect to have any element of maximum demand charging for power used at night? Discuss your findings and whether there is a realistic way in which the electricity costs could be reduced.
  Try
to form some sort of assessment of load factor for the establishment. What order of load factor would you expect in your home? There are considerable cost disadvantages for generators and distributors of electricity in having customers with low load factors. Why is this?

Effect on electricity charges of improving load factor

As an example, suppose a factory has an energy requirement of 750 000 kWh/annum. The electricity tariff has a maximum demand charge of £18/kW. Let us consider the effect of changing load factors on the maximum demand charge.

(a)   With a load factor of 0.25

$$0.25 = \frac{750\,000}{\text{Maximum demand} \times 365 \times 24}$$

$$\text{Maximum demand} = \frac{750\,000}{0.25 \times 365 \times 24} = 342.47 \text{ kW}$$

Annual maximum demand charge $= 342.47 \times £18 = £6164.46$

(b)   With a load factor 0.75, using a calculation similar to that above, the maximum demand reduces to 114.16 kW.

$$\text{Maximum demand charge} = 114.16 \times £18 = £2054.88$$

$$\text{Saving} = £4109.58$$

It can be seen therefore that an improvement in load factor leads to a cost saving where the tariff includes a maximum demand element.

This improvement in load factor may be achieved in a number of ways. Suppose a factory with a metal foundry and a machine shop both started up simultaneously in the morning. There will be a heavy demand for the first hour or so after which the demand will reduce as furnaces reach working temperature, motors warm up and in winter, the workshops reach a comfortable temperature and possibly heating and lighting are switched off. The load factor is low since the heavy demand only lasts for a small proportion of the working day.

By either working a night shift in the foundry or starting it up, say, two hours earlier than the machine shop, the demand by the foundry can be reduced to a low value when the machine shop starts running. The maximum demand is thereby reduced and the load factor increased. The total energy remains at about the same level.

The load factor in commercial premises could be improved by using off-peak storage heating instead of direct-acting equipment so that the heating load is off when lighting and other services are required for a morning start.

Increasing the number of hours that a particular piece of equipment works improves its load factor. Where several pieces of equipment are used only intermittently, ensuring that only one is used at a time improves load factor.

Effect of power factor improvement

Where a tariff incorporates a maximum demand charge based on kVA or has a power factor penalty clause as in Example 8.2, savings can be achieved by improving a low power factor. Such a tariff seeks to discourage low power factor since this involves larger currents than necessary to perform a given amount of work. Conductors, transformers and switchgear must therefore be larger than for the same load at unity power factor. Larger equipment means higher capital charges and higher currents mean increased losses all of which have to be paid for. As an alternative to charging for kVA of maximum demand, $kVA_r h$ (reactive volt-ampere hours) may be measured and charged for as well as kWh.

Consider a factory with a maximum demand of 400 kW paying for electricity on a tariff of £15 per kVA of maximum demand.

(a) At a power factor of 0.2 lagging. $\phi_1 = \text{INV. Cos} \ (\cos^{-1}\ 0.2 = 78.46°$ (see Figure 8.4).

$$kVA_1 = \frac{400}{0.2} = 2000$$

Maximum demand charge = £15 × 2000 = £30 000.

(b) At a power factor of 0.6 lagging. $\phi_2 = 53.1°$.

$$kVA_2 = \frac{400}{0.6} = 666.67$$

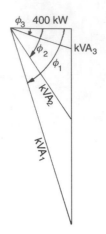

Figure 8.4

Maximum demand charge = £15 × 666.67 = £10 000.

(c)   At a power factor of 0.9 lagging. $\phi_3 = 25.8°$.

$$kVA_3 = \frac{400}{0.9} = 444.4$$

Maximum demand charge = £15 × 444.4 = £6666.

(d)   At unity power factor $\phi = 0°$ and kW = kVA.
       Maximum demand charge = £15 × 400 = £6000.

Thus considerable savings can be made by improving the operating power factor. The increments of saving get less as the power factor approaches unity. From 0.2 to 0.4 gives a greater saving than from 0.8 to unity for example.

A capacitor draws a current which leads on the supply voltage by 90° and therefore has no active component. It is totally reactive and multiplying by the supply voltage V gives VI_C, leading volt-amperes reactive. (See Figures 8.5 and 8.6.)

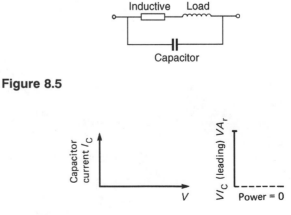

Figure 8.5

Figure 8.6

Adding the capacitor to the load does not affect the load power factor but since the leading and lagging volt-amperes reactive are in direct phase opposition the arithmetic difference may be taken as shown in Figure 8.7.

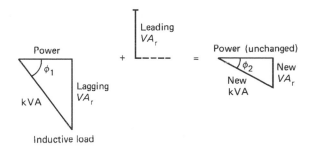

Figure 8.7

New value of VA_r = Lagging VA_r – Leading VA_r

The overall phase angle has been reduced from ϕ_1 to ϕ_2 and the value of kVA has been reduced.

As well as capacitors, synchronous motors, which may be used to drive loads within a factory or may just run light, can be made to operate with a leading power factor by varying their excitation (field) current so achieving the same degree of compensation. Synchronous machines may operate as generators when driven or as motors when they draw a current from an electrical supply.

Example 8.4

An industrial load has a maximum demand of 300 kW at a power factor of 0.6 lagging. Calculate the saving in maximum demand charges and the overall saving if a capacitor is fitted which draws 150 kVA$_r$. The tariff is £18/kVA of maximum demand. The annual capital charges for the capacitor are £850.

(Note that capacitors, like all plant, have to pay capital charges as already discussed. They may be bought or rented; either way there is a charge to be met. For a capacitor the running charges are considered to be zero since there is virtually no power loss.)

$\cos \phi = 0.6$ Therefore $\phi = \text{INV} \cos 0.6 = 53.13°$

$$\text{kVA} = \frac{300}{0.6} = 500$$

$$\frac{\text{kVA}_r}{300} = \tan \phi = \tan 53.13° = 1.33$$

$$\therefore \text{kVA}_r = 300 \times 1.33 = 400$$

or by Pythagoras' theorem: $\text{kVA}_r{}^2 = \text{kVA}^2 - \text{kW}^2$.

Maximum demand charge = £18 × 500 = £9000.

Test your knowledge 8.4

Calculate the saving in maximum demand charge if a factory with a maximum demand of 500 kW at a power factor of 0.7 lagging improves this to 0.9 lagging. The maximum demand charge is £20/kVA.

What is the overall saving if each kVA$_r$ of compensating equipment has an annual charge of £6?

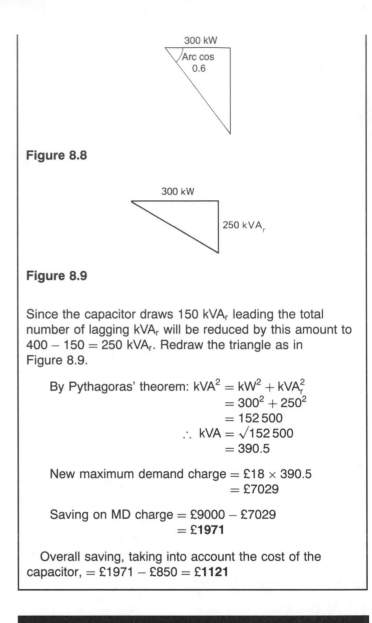

Figure 8.8

Figure 8.9

Since the capacitor draws 150 kVA$_r$ leading the total number of lagging kVA$_r$ will be reduced by this amount to $400 - 150 = 250$ kVA$_r$. Redraw the triangle as in Figure 8.9.

$$\text{By Pythagoras' theorem: } kVA^2 = kW^2 + kVA_r^2$$
$$= 300^2 + 250^2$$
$$= 152\,500$$
$$\therefore \ kVA = \sqrt{152\,500}$$
$$= 390.5$$

$$\text{New maximum demand charge} = £18 \times 390.5$$
$$= £7029$$

$$\text{Saving on MD charge} = £9000 - £7029$$
$$= \mathbf{£1971}$$

Overall saving, taking into account the cost of the capacitor, $= £1971 - £850 = \mathbf{£1121}$

Activity 8.2

Measure the effectiveness of power-factor correction capacitors. The test circuit is as shown in Figure 8.5. The inductive load is either an inductive loading bank or an inductor in series with a rheostat. The capacitive compensation is either a loading bank or several discrete mains-type capacitors. An ammeter is included in each leg and a third in the main supply lead together with a wattmeter. Connect the circuit to a suitable alternating supply, either at mains voltage or if more appropriate to a reduced-voltage supply, connecting a voltmeter across the supply terminals.

With no capacitance in circuit measure the supply current. Determine the power factor using the VA:power relationship. If possible adjust the inductive circuit to give a low power factor. Add capacitance in steps noting the increasing capacitive current and the reduced overall current. Comment on the change in overall power factor as the capacitance is increased. Does there come a time when adding capacitance increases the overall current? Why is this? Does the total circuit power change during the test? If so why might this be? Why would power factor correction not be appropriate for a domestic consumer?

It is usually uneconomic to correct the power factor to unity since the costs of so doing become greater than the savings as the power factor approaches this value.

Consider the case of a factory with a maximum demand of 200 kW at a power factor such that it requires 300 kVA, lagging. This is condition (1) in Figure 8.10.

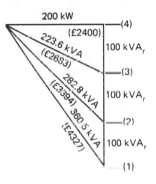

Figure 8.10

We will consider the effects of adding capacitor banks to the system in three stages each drawing 100 kVA$_r$ leading. The capital charges on each bank are £650 and the maximum demand charge is £18/kVA.

With no capacitors in circuit the maximum demand is 360.5 kVA and the maximum demand charge is £18 × 360.5 = £6489.

Adding one bank of capacitors improves the power factor so reducing the kVA demand to 282.8. MD charge = £18 × 282.8 = £5090 (point (2) in Figure 8.10).

Adding the next bank of capacitors reduces the kVA demand to 223.6 giving an MD charge of £4025 (Point (3) in Figure 8.10).

Adding the final capacitor bank brings the power factor to unity when the kVA demand is equal to the number of kW and the maximum demand charge is £18 × 200 = £3600 (Point (4) in Figure 8.10).

The savings from Point (1) to Point (2) = £1399 at a cost of £650 spent on capacitors.

Net saving = £749

From point (1) to Point (3) the savings = £2464 at a cost of £1300.

Net saving = £1164

Correcting to unity, from Point (1) to Point (4) the saving in maximum demand charge = £2889 at a cost of £1950.

Net saving = £939

Alternatively consider changing from Point (3) to Point (4). The saving in maximum demand charge = £4025 − £3600 = £425 and this has cost £650 in capacitor charges.

Figure 8.11 shows a graph of savings against power factors for the above case. The maximum savings are achieved by correcting to a power factor of about 0.94. Going beyond this value costs more for capacitors than can be saved on maximum demand charges. Different capacitor and maximum demand charges will alter the power factor for maximum savings.

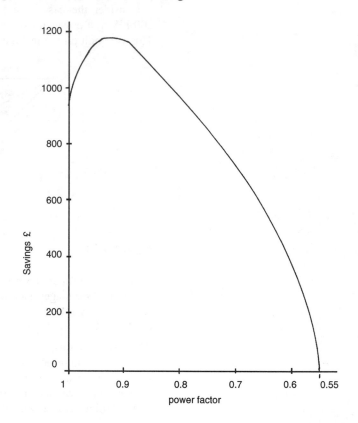

Figure 8.11

Problems on tariffs and power factor correction

1   The cost of generation and transmission of electrical energy is divided into two parts. Name and briefly explain these two components of cost.

2   Define the term 'maximum demand'. Why is it relevant to electricity supply?

3   Why is the domestic consumer not charged on a tariff based on maximum demand?

4   Draw a triangle for a lagging power factor load showing watts, volt-amperes and volt-amperes reactive. Define the power factor of the circuit in terms of these quantities.

5   Define the term 'diversity factor'. Why is it important when designing an electricity system?

6   Explain how it is that improving the load factor on a piece of equipment can lower the average cost per kWh consumed.

7   An industrial load draws a power of 75 kW and 55 kVA, lagging. What is the power factor of the load?

[0.806]

8   A factory has a maximum demand of 200 kW and uses 500 000 kWh per year. What is the value of the load factor?

[0.285]

9   Where a factory has a large consumption, the choice of taking a bulk supply at high voltage may be offered. Why would this be likely to result in a lower rate tariff being available?

10   A private house contains the following equipment:

Electric lamps, total power 1100 W
Television set 200 W
Electric iron 750 W
Two electric fires, total 4000 W
Washing machine with heater 3000 W
Food mixer 500 W
Electric cooker 10 000 W
Electric kettle 3000 W

(a)   Calculate the diversity factor when (i) 200 W of lighting, the television set and both electric fires are working, (ii) the cooker is operating at one half of maximum power, the electric iron and kettle are switched on.

(b)   Calculate the load factor for the installation if in one quarter 1250 kWh are consumed and the maximum demand was 50 per cent of the total equipment listed above.

(c)   Calculate the total electricity charge for the quarter and the average cost per kWh for a tariff: Fixed charge £9.5 + 8.5p/kWh of energy consumed.

[(a) (i) 0.195 (ii) 0.388, (b) 0.05, (c) Total £115.75, average 9.26p/kWh]

11   During eight successive half hours the energy meter connected to the supply lines to a commercial premises was noted to make the following advances: 100, 150, 125, 175, 200, 150, 125 and 130 kWh respectively. Assuming that these are the highest noted during the year, determine the maximum demand charge on a tariff of £20/kW of maximum demand.

[£8000]

12   Why is there no incentive for domestic consumers to practise (a) power factor improvement or (b) load factor improvement?

13 Calculate the values of the quantities missing from the following table.

	kW	kVA	kVA$_r$	power factor
(a)	150	?	75	?
(b)	150	?	?	0.65
(c)	?	?	300	0.7

[(a) 167.7 kVA 0.894 p.f., (b) 230.77 kVA 175.4 kVA$_r$, (c) 294.06 kW 420.1 kVA.]

14 The half-hourly advances of a kWh meter are as follows during a working period of 8 hours: 100, 125, 175, 200, 300, 500, 450, 400, 350, 350, 450, 500, 490, 502, 350, 150. These readings contain the highest values obtained throughout the year. The load factor over the year is 0.2. Assuming the power factor to be 0.8 lagging, calculate the annual cost of electricity on two tariffs: (a) £17/kVA of maximum demand +4p/kWh; (b) £22/kW of maximum demand +4.3p/kWh.

[MD = 1004 kW, (a) MD = 1255 kVA kWh consumed = 1 759 008 cost = £91 695, (b) Cost = £97 725]

15 A factory draws a load of 250 kW at a power factor of 0.65 lagging. Calculate: (a) the number of (i) kVA and (ii) kVA$_r$ drawn by the load;
(b) the rating of a capacitor to be connected across the input terminals to the factory which would cause the overall power factor to be 0.95 lagging.

[(a) (i) 384.6 kVA (ii) 292.3 kVA$_r$, (b) 210 kVA$_r$]

16 Calculate the rating of a capacitor which would be needed to correct the power factor of an industrial load of 650 kVA at a power factor of 0.7 lagging to 0.9 lagging.

[243.8 kVA$_r$]

17 A factory with a foundry and a machine shop has a maximum demand of 750 kW and a load factor of 0.2. The electricity tariff is £20/kW of maximum demand +6p/kWh of energy consumed.
(a) Calculate the annual cost of electricity on this tariff.
(b) To save money a two-shift system of working is introduced. The foundry starts work earlier in the morning whilst the machine shop commences rather later in the day. By this means the maximum demand charge is reduced since all the equipment will not be switched on simultaneously at the beginning of each day. The working pattern increases the number of kWh consumed by 5 per cent but increases the load factor to 0.35. Calculate the saving in electricity charges.

[(a) Total cost = £93 840, (b) New MD = 450 kW, total cost = £91 782; saving £2058]

18 An industrial load has a maximum demand of 500 kW at a power factor of 0.65 lagging. Calculate the saving in maximum demand charges and the overall saving if a capacitor bank is fitted which draws 150 kVA$_r$. The tariff is

£15/kVA of maximum demand. The annual charges for the capacitor are £600.

[MD saving = £1601.4, Overall saving £1001.4]

19 A factory uses 850 000 kWh per annum. The load factor = 0.3. The tariff is £20/kVA of maximum demand +4p/kWh. Calculate: (a) the maximum demand in kW and kVA assuming a power factor of 0.7 lagging and (b) the new MD in kVA and the size of capacitor required (in kVA$_r$) if the electricity bill is to be reduced by £2000 per annum.

[(a) 323.44 kW, 462.06 kVA, (b) to save £2000, kVA = 362.06. Required kVA$_r$ = 167.3]

20 A factory consumes 1.2 million kWh per annum. It has a load factor of 45 per cent and operates at an average power factor of 0.65 lagging.
(a) Calculate the total electricity charges on a tariff of £18/kVA of maximum demand +5.5p/kWh.
(b) Calculate the savings to be made by improving the power factor to 0.9 lagging. The capital charges on power factor correcting equipment are £7 per kVA$_r$.

[(a) Charges = £8429.94 + £66 000 = £74 430, (b) new MD charge = £6087.96 Save £2342. Capacitor costs £1459.3, Overall save £882.7]

21 For the month of July an industrial user operates on the following tariff:

Availability charge £1.20 per kVA
Reactive charge for each kVA$_r$h in excess of half of the total number of kWh consumed in the month = 0.4p
Standing charge = £25
Metering charge = £10
Units consumed during the day = 4.5p/kWh
Units used at night = 2.5p/kWh.

The user has declared an availability requirement of 150 kVA. Total energy usage = 8900 kWh of which 2100 kWh were used on the night tariff.
Reactive usage = 6000 kVA$_r$ h.
Determine the total cost of energy for the month and the average cost per kWh.

[Costs = £(180 + 6.20 + 10 + 25 + 306 + 52.50) = £579.7. Average = 6.51p/kWh]

22 A workshop uses electricity charged on the following tariff for one winter month of 30 days:

Maximum demand charge £7/kW
Metering charge £6
Units charge at 5.5p/kWh for the first 150 kWh/kW of maximum demand
All other energy at 4p/kWh.

(a) Calculate the average cost per kWh of energy if the maximum demand is 20 kW given that the load factor for the workshop is 20 per cent.

(b) If the same energy requirement is maintained by an improvement in working practices leading to an increased load factor to 30 per cent determine (i) the new maximum demand and (ii) the new average cost per kWh.

[(a) £(140 + 6 + 2880 × 5.5p) = £304.4. Average 10.57p/kWh, (b) New MD = 13.33 kW, Cost £(93.3 + 6 + 110 + 35.2) = £244.51. Average 8.49p/kWh]

Multiple choice questions

1 The capital charges involved in building a power station and associated transmission lines are made up of:

A interest paid on capital borrowed or dividends
B cost of fuel
C wages paid to operatives
D depreciation allowance towards the provision of new plant
E cost of repairs to the plant
F cost of system losses.

Select *two* of the six alternatives.

2 The power in a single-phase circuit operating at 250 V is 1250 W. The power factor is 0.6 lagging. The value of the reactive current I_r is:

A 5 A
B 8.33 A
C 6.67 A
D 4 A.

3 A factory has an energy meter used for the determination of the maximum demand charge. The readings for the six hours containing the maximum for the year were: 150, 175, 200, 201, 202, 198 kWh. The power factor has an average value of 0.75 lagging. The tariff is £20/kW of maximum demand. The MD charge for the year will be:

A £5386.66
B £4040
C £8080
D £10 773.33.

4 The load factor of an industrial load is 0.25. One million kWh of energy are consumed during one year. The average power factor is 0.7 lagging. The maximum demand of the load is:

A 319.63 kW
B 456.62 kW
C 652.3 kW

5 Improving the load factor of a factory can reduce the electricity bill because:

    A it reduces the maximum demand for a given energy
consumption

    B it increases the maximum demand

    C it improves the power factor

    D it reduces the energy consumption for a given maximum
demand.

6 A factory with a maximum demand of 200 kW has a load
factor of 0.3. The energy consumed during a period of 90
days is:

    A 129 600 kWh

    B 60 kWh

    C 1.44 million kWh.

7 A factory with a maximum demand of 550 kW at 0.65 power
factor lagging improves its power factor to 0.95 lagging by the
use of capacitors. The tariff is £16/kVA of maximum demand
per annum. The annual saving in electricity charges is:

    A £7398

    B £2890

    C £4257.

8 Domestic electricity is paid for on one of the following tariffs.
Which is it?

    A maximum demand in kW + charge per kWh

    B maximum demand in kVA + charge per kWh

    C fixed charge + charge per kWh

    D charge per kWh + penalty for poor power factor.

9 A consumer with a poor power factor charged on a tariff with
a maximum demand charge per kW should be A or B his
power factor since by so doing he will C or D.
Select statements from A to D below to make a correct
statement.

    A encouraged to improve

    B discouraged from improving

    C reduce substantially the number of kW of maximum
demand and so reduce his electricity bill

    D spend money unnecessarily since it will have no effect
on his electricity bill.

10 A factory with a maximum demand of 350 kVA at 0.7 power
factor lagging installs a capacitor bank which draws 150 kVA_r
leading. The value of the kVA drawn by the factory after
adding the capacitor is:

    A 249.9 kVA

    B 245 kVA

    C 265 kVA.

11 A factory wishes to expand production. This may be achieved
by either (a) increasing the amount of production plant
installed or (b) by working existing plant for two shifts instead
of one as at present.

A saving in the electricity bill will be achieved by scheme (b) as opposed to scheme (a) since:

A   the power factor will be improved
B   the load factor will be improved
C   the diversity factor will be improved
D   the actual energy charge (kWh consumed) will be reduced
E   the MD charge will be reduced.

9 Regulations

Summary

This chapter explains the need for and discusses the scope of the Electricity Supply Regulations, the Regulations for Electrical Installations, the Health and Safety at Work Act and the Electricity at Work Regulations.

Safety and regulations

Inadequate control of electricity can give rise to serious dangers due to fire and shock. In the early days of the electricity public supply system the powers involved were small and faults were frequent, often resulting in whole plants being shut down. At least one power station was destroyed by fire when a fault developed and was not cleared. With the vast amounts of power involved today such a situation cannot be allowed to develop and over the years regulations applying to suppliers and consumers have been drawn up to prevent serious situations and accidents as far as is reasonably practicable. A major accident causes the regulations to be reviewed to see how best to prevent a future similar occurrence but it must be said that many of the original regulations were so well and widely cast that remarkably few changes have been required over the years.

These regulations are:

The Electricity Supply Regulations 1988;
The Institution of Electrical Engineers' Regulations for Electrical Installations.

In addition all workers are covered by *the Health and Safety at Work Act*. Under this Act *the Electricity at Work Regulations 1989* have been brought into force.

There are also special regulations applying to coal mines, quarries, cinemas, places of public entertainment and oil refineries.

The Electricity Supply Regulations

These are administered by the Engineering Inspectorate of the Electricity Division of the Department of Energy. They are designed to secure the safety of the public and to ensure a proper and safe supply of electrical energy and apply to the electrical distribution system covering construction, operation, protection and earthing, maintenance and safety up to the consumers' terminals.

These regulations may apply additionally to those imposed by the Electricity at Work Regulations.

Certain sections relate to consumers' installations in as much as they give the Supply Companies powers to insist upon certain standards of work before the supply can be connected. The standards generally required are those laid down in the IEE Regulations for Electrical Installations. In addition, a supply may be refused to a consumer whose equipment is likely to affect other consumers adversely.

The Supply Companies have to declare the voltage of the supply and maintain this within 6 per cent of that value.

The Generating Companies maintain the frequency within specified limits and maintain an average of 50 cycles per second (50 hertz) over 24 hours (so keeping electric clocks correct).

Both Generating and Supply Companies have very strict sets of safety rules and operate a 'Permit-to-work' system which ensures an extremely low accident rate.

The Institution of Electrical Engineers' Regulations for Electrical Installations

The overall objective of these regulations is to ensure the protection of people and livestock from fire, shock or burns from any installation that complies with their requirements.

The current edition of these regulations is the sixteenth. This was published in May 1991 and came into force on 1 January 1993. Since October 1992 the IEE Regulations have become a British Standard: BS 7671, 1992 Requirements for Electrical Installations.

The regulations are based on the international regulations drawn up by the International Electrotechnical Commission with the aim of having a common set of regulations for all EU countries. (World consensus on electrical safety is a desirable goal but probably not yet practicable.)

The 15th Regulations were complete within themselves and included much guidance material. The 16th edition omits the guidance material; only the legal skeleton is present. There are eight sets of guidance notes to help in their interpretation and aid good practice. These cover broadly the material in the guidance notes of the 15th edition.

The regulations covering a particular aspect of working, for example earthing or circuit protection, are almost invariably subdivided and are to be found in different parts of the Regulations. The overall picture can only be obtained after a consideration of all the relevant sections, which involves much use of the index. For this reason, only a very limited number of topics are quoted here with a selection of locations where references may be found.

- Electrical conductors shall be of sufficient size and current-carrying capacity for their intended purpose. [434-03-03]
  (The location quoted breaks down into: Part 4, Chapter 3, Section 4, Group or Subsection 3, the 3rd Regulation)
- All conductors shall be suitably insulated or protected to prevent danger, bearing in mind the environment in which they are working. [522-01 to 522-05 and parts of 523 and 527]
- Circuits shall be protected against overcurrents. [432-02, 432-03 and parts of 433 and 473]

- Circuits must be adequately protected against fault currents. [432-02, 432-04 and elsewhere]
- The requirements for equipment to be adequately earthed and the methods appear in Sections 130-04, 542-04 and others.
- Requirements for installation of lighting circuits appear in Sections 422-01, 553-03 and others.

In Part 4, tables of cable sizes for different currents are provided together with factors to be applied where the cables operate in a high ambient temperature or where several cables are grouped together, possibly within a conduit. (See also Chapter 5, 'Permitted volt drops in cables'.) Where cables are to be run in spaces where thermal insulation is present, such as within a thermally insulated wall, again there are further factors to be taken into account.

Different factors have to be used according to whether the circuit is protected by an HRC fuse, a miniature circuit breaker or a rewireable fuse. Where a rewireable fuse is used, allowance must be made for the fact that it is able to carry a large overcurrent for a considerable time before clearing. For this reason no rewireable fuse can have a rating greater than 0.725 times that of the smallest cable it is protecting. [433-02] (See also Chapter 7, Activity 7.1.)

Compliance with the IEE Regulations is implicitly required by The Electricity at Work Regulations (1989) so that should an accident occur due to non-compliance with these regulations, legal proceedings could result.

Certain installations where there are exceptional risks must also comply with additional Statutory Regulations, namely those in cinemas, coal mines, quarries, oil refineries and horticultural and agricultural installations. Work conforming to the IEE Regulations will generally satisfy these Statutory Regulations in so far as they relate specifically to the associated buildings. Where these Statutory Regulations have different requirements from or in addition to those of the IEE Regulations, the former requirements take precedence. Installations of a special character will require the additional advice of a suitably qualified specialist. The Regulations recognize that not all eventualities are covered.

The Health and Safety at Work Act

The Health and Safety at Work Act (HASAWA) came into force in 1974. It was introduced as a result of a report by the Robens Committee on Health and Safety, one of whose findings was that a considerable number of accidents were the result of apathy on the part of workers and management alike.

Up to this time there had been a number of Factories Acts, the first being in 1833 which limited the working hours of children. All these Acts were aimed at premises, e.g. offices, workshops and railway premises. They aimed at making these safer places to work in. The Health and Safety at Work Act, however, is directed at people and activities. The underlying objective is to involve everybody at the workplace, whether this be in a factory or on a farm, at a workbench or driving a lorry, and to create an awareness of the importance of achieving high standards of health and safety. This gives far greater coverage than the old Factories Acts, protecting everybody at work (except domestic servants employed in private households). It also extends the protection to the general public and to visitors to places of activity who are not employees.

The Health and Safety at Work Act is an 'Enabling Act'. This means that it does not itself contain a lot of detailed regulations but enables other, often older acts, such as The Offices, Shops and Railway Premises Act and the more recent Electricity at Work Regulations (1989), to be used to cover most working situations. Examples will serve to demonstrate the point. Previous to the HASAWA there was no specific Act for schools and colleges, so that, with a few exceptions, the staff, pupils and students were not covered by legislation. The school secretary, recognized as an office worker, was protected under the Offices, Shops and Railway Premises Act and could legally demand that the office temperature be maintained at a satisfactory level, whereas teaching colleagues in an adjacent room were not so covered. In a college laboratory a pressure vessel or boiler not originally covered by legislation would now be equated to its equivalent situated in a factory and would therefore need to be inspected and tested in the same way.

Outline of the requirements of the HASAWA

1   The employer must ensure as far as is reasonably practicable, the health, safety and welfare of all his employees by the provision of:

    (a)   safe plant and equipment
    (b)   safe materials handling, storage and transport facilities
    (c)   safe systems of work, with adequate training and supervision
    (d)   safe premises and working environment.

    In certain circumstances he must provide for periodic medical examination of his employees. This will be particularly relevant where toxic substances are being handled, for example.
    He must ensure that no process in the workplace yields harmful emissions which would damage the health of employees or the general public.
    A copy of the firm's safety policy must be available to all employees. This lays down procedures for safe working, and who is responsible for carrying out the policy. This person must be someone with executive power.

2   Employees must take reasonable care of their own health. They must, where relevant:

    (a)   wear safety glasses, ear muffs and protective clothing
    (b)   cooperate with colleagues and supervisors to promote safe working.

The situation where a machine guard has been removed in order to clear an obstruction and not subsequently replaced, or removed to facilitate extra production, might be envisaged. The person removing the guard, any person using the machine in an unguarded state and the supervisor who should see that the guards are maintained in a working condition could all be deemed negligent.

Manufacturers, suppliers and importers have to ensure that goods and equipment supplied are safe for use in the manner for which they are designed. This responsibility can be traced back to the designer of the equipment if necessary. Equipment manufactured and tested to the levels set by the relevant British Standard will satisfy this requirement in every way. If the material is imported with no British Standard number it must have a 'Harmonization Document' which will guarantee that it has been designed and manufactured to a standard at least as good as the British Standard.

All substances must be adequately labelled to ensure that they will be correctly used and, particularly when they are being transported in bulk, to enable the emergency services to deal quickly and correctly with any spillage.

The Health and Safety Inspectorate

The enforcing body of the HASAWA is the Health and Safety Executive which includes Her Majesty's Factory Inspectorate. This body has overall responsibility for the day-to-day operation of the Act and its enforcement. The factory inspectorate is now divided into Area offices and further into Industry Groups to provide expert inspectors for particular industries. However, the scope of the Act is much wider than industry, and under the 'Enforcing Authority Regulations' local authorities are responsible for its enforcement in certain areas of its application. For instance, the County or Town Councils oversee offices to see that they comply with the Offices, Shops and Railway Premises Act; the Local Education Authorities oversee schools and colleges of further education.

The inspectors have wide powers. These include the right to inspect and investigate anything in the course of their duties. They may enter premises at any 'reasonable time', or at any time at all if they have reason to suppose that a dangerous situation exists within. If the inspector thinks that his entry may be impeded, he may bring a police officer with him. If the situation warrants it he may issue either an 'Improvement Notice' or a 'Prohibition Notice'. The Improvement Notice will require a situation or process to be improved to decrease the risk of danger. The improvement will have to be carried out within a specified period of time. The recipient of the notice may appeal, but this appeal does not increase the length of time allowed. Failure in the appeal means that the improvement must be carried out by the original date. A Prohibition Notice means that the process must stop immediately. It can only restart when the process has been made safe and has been approved by the Inspectorate.

The Electricity at Work Regulations (1989)

This set of regulations has been made under the HASAWA. They have been drawn up to ensure safe working with electricity in the workplace and aim to prevent unsafe installations and practices which otherwise might be used out of ignorance or to save costs. There are 33 regulations in all; 16 general regulations followed by 12 relating to mines, followed by regulations concerned with Exemptions, Extension outside Britain and Modifications. There are distinct similarities between the general regulations and the IEE Regulations. This is not surprising, since safe use of electricity at work and safe installation practice are inevitably closely linked. Since the regulations are designed to promote safe working, it follows that non-compliance will create a hazardous or unsafe condition.

Enforcement of these regulations under the HASAWA is by the factory inspectorate.

We will examine some of the general regulations:

- **Reg 3.** Employers, self-employed persons and managers of mines or quarries must comply with the provisions in these regulations. In addition it is the duty of every employee to cooperate with his employer in complying. This makes it equally the responsibility of supervisor and supervised to ensure safety. This is a reinforcement of Section 7(b) of the HASAWA; every person must take responsibility in association with others for his own safety. In addition he must not take any action which either directly or indirectly will put anybody else at risk.
- **Reg 4.** All systems must be of such a construction and be maintained in such a condition as to prevent danger. Equipment provided for the protection of persons working on or near electrical equipment must be suitable for that use and maintained in good condition.
- **Reg 6.** Electrical equipment which may be subjected to adverse conditions of weather, heat, chemical attack etc. must be of such a construction as to withstand such conditions.
- **Reg 8.** Any conductor which can become charged as a result of the method of usage or of a fault must have a suitable means of preventing danger to life by contact with that conductor. This will often be by means of earthing (the metal case of a motor, for example).
- **Reg 10.** Every joint or connection must be mechanically and electrically suitable for the situation. This means, for example, that when making an earth connection, the wire should be held mechanically, perhaps with a screw or twisted joint rather than just attached by soft solder which might fail under mechanical movement.
- **Reg 11.** Efficient means must be provided for protecting a system from excess current. Often this is a fuse or circuit breaker actuated by an excess current detection device.
- **Reg 12.** Suitable identification of circuits must be provided together with means of isolating them from the electricity supply.
- **Reg 13.** When equipment has been made dead for work to be carried out, adequate precautions must be taken to ensure that the equipment cannot be re-energized during that work.

- **Reg 15.** Adequate working space and means of access must be provided to safely carry out work on equipment.
- **Reg 16.** Every person carrying out work on equipment must be in possession of such technical knowledge or experience as is necessary to prevent danger or be under the direct supervision of a person so qualified.

Comparison between the scopes of the regulations

The Supply Regulations apply up to the supply point in the factory or premises. The IEE Regulations relate to installation practice on consumers' premises. The IEE Regulations is a code of practice and a British Standard which is widely accepted in the UK and compliance with these Regulations is likely to satisfy the Electricity at Work Regulations 1989 and the Supply Regulations in as far as these affect consumers.

The Electricity at Work Regulations have the force of law behind them. Prosecution can result from non-compliance and the IEE Regulations could be quoted to demonstrate malpractice.

The IEE Regulations are likely to be the basis of a contract between an installation firm and the purchaser of an installation.

Activity 9.1

Regulation 13 of the Electricity at Work Regulations requires that precautions shall be taken to prevent electrical equipment which has been made dead in order to prevent danger whilst work is being carried out on or near that equipment, from becoming electrically charged during that work if danger may thereby arise. This will probably involve use of the means required in Regulation 12. Read Regulation 12 in full. One way of complying with Regulation 13 is to use a 'Permit-to-Work' system. Such a system is used in power stations and by the supply companies. Investigate and report on how this, or any other suitable system, is operated. (Work on high-voltage equipment and permits to work are detailed in BS 5405 with updates in BS 6423 and 6626.)

Activity 9.2

Both the IEE Regulations and the Electricity at Work Regulations pay great attention to system earthing. Investigate this requirement in both sets of Regulations and report on why earthing is required and how it is achieved. Why do some domestic premises have an earth electrode whilst others do not? What is a PME system? In what circumstances is it necessary to bond electricity, water and gas systems?

Activity 9.3

The HASAWA requires certain things of both employers and employees. Detail what these requirements are and discuss one such requirement in detail as it affects you at your place of work.

Problems on regulations

1   What is the basic difference in approach of the Health and Safety at Work Act as compared with the older Factories Acts?

2   What is the general objective of all electricity supply and wiring regulations?

3   What powers do the Electricity Supply Regulations give to the Supply Companies with respect to consumers' equipment?

4   What do the Electricity Supply Regulations have to say about terminal voltage at the consumer's terminals?

5   What are the powers of the Factory Inspectorate with respect to the HASAWA?

6   What is the general aim of the IEE Regulations for Electrical Installations?

7   Under what circumstances may additional requirements to those laid down in the IEE Regulations have to be met?

8   What requirement with regard to frequency is normally observed by the generating companies?

9   What powers do Factory Inspectors have with regard to unsafe equipment or working practices? How might they initiate corrective measures?

10   What does the HASAWA have to say about the labelling of substances?

11   In the context of the HASAWA what is (a) an Improvement Notice and (b) a Prohibition Notice?

12   Which of the regulations discussed in this chapter relate to (a) the consumers of electrical energy and (b) the suppliers of electrical energy?

13   What is a 'Harmonization Document'?

Multiple choice questions

1   The IEE Regulations detail the requirements in respect of the following:

A   frequency stabilization at an average of 50 cycles per second
B   protective earthing and the safety of electrical installations
C   voltage stabilization at ± 6 per cent of declared value
D   safety of supply cabling to premises
E   access to working spaces
F   the safe working of plant in factories
G   ensuring safety from fire and shock in the utilization of electricity in and around buildings.

Select TWO of the choices A to G.

2   There are three sets of Regulations governing the supply and utilization of electrical energy. These are:

A   The Electricity Supply Regulations
B   The Electricity at Work Regulations
C   The IEE Regulations.

These Regulations cover the following areas of application:

(a)   installation of equipment in buildings
(b)   safe working practices and operation in workshops and the like
(c)   equipment up to the supply point in a factory or premise.

Link the particular regulation to its application. (Example: A (a); B (c) etc.)

3   The Health and Safety at Work Act is an enabling act. It enables:

A   the Supply Regulations to be extended to cover consumers' premises
B   the IEE Regulations to be enforced at law
C   the Factories Acts to be applied to like installations on a farm or in a college workshop.

4   The Electricity Supply Regulations are designed to secure the safety of the public. There are clauses relating to:

A   the state of the consumer's installation
B   suitable means for cutting off the supply to a piece of faulty equipment
C   earthing the low voltage side of a step-down transformer to prevent shock
D   the supply voltage to the consumer
E   efficient means of ventilating industrial premises.

Select TWO of the above statements.

5   The Electricity at Work Regulations require that steps be taken to prevent electric shock. Measures to be taken include:

A   double insulation
B   earthing exposed metal work
C   bonding of services (water, gas and electricity)
D   earthing one phase of the supply
E   operating at low voltages
F   earthing more than one point on the supply transformer
G   use of a suitable protective device such as a fuse, bearing in mind the fault level.

TWO of the above statements are NOT correct. Select these two.

6   A 'Harmonization Document' is required when:

A   price equality is required between electrical equipment bought in the UK and in the other EU countries

B   importing foreign-made electrical equipment so ensuring that it complies with, or exceeds, British Standards

C   buying cables from various manufacturers to ensure consistent cross-sectional area and current-carrying capacity

D   importing CDs and audio tapes to ensure that they can be played on UK-manufactured equipment.

Appendix to Chapter 1 The three-phase delta connection

The voltage phasors for a three-phase, star-connected system as associated with a generator or supply transformer are shown in Figure A.1.

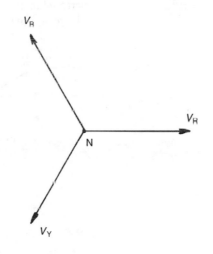

Figure A.1

The red, yellow and blue phasors rotate in an anticlockwise direction in what is called the *positive phase sequence*. The neutral is at earth potential and the phase voltages are measured from the supply lines to the neutral. To determine the magnitude and phase of the voltages between lines the phasor subtraction as shown in Figure A.2 is performed.

The potential difference between the red line and the yellow line is the difference between their respective values to earth or neutral. (If the voltages concerned were direct, if the red line was at 200 V above earth and the yellow line were, say, 50 V above earth, the

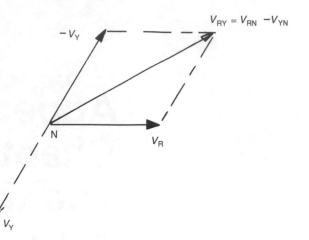

Figure A.2

potential difference between the two lines would be $200 - 50 = 150$ V.

We require to evaluate $V_R - V_Y$. $(-V_Y)$ is the V_Y phasor reversed. In Figure A.2 the addition $V_R + (-V_Y)$ is performed and the result is V_{RY}, the line voltage between red and yellow lines. By scaling, or by resolution into vertical and horizontal components, it is found that $V_{RY} = \sqrt{3}V_R = \sqrt{3}V_Y$ or in general $V_{LINE} = \sqrt{3}V_{PHASE}$. By performing the phasor subtraction in turn for the other phases, Figure A.3 may be constructed.

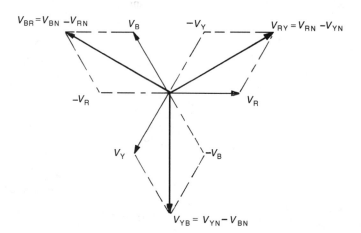

Figure A.3

In Figure A.3, $V_{YB} = V_{YN} - V_{BN}$ and $V_{BR} = V_{BN} - V_{RN}$.

The three line voltages are displaced each from the other by $120°$, all are equal in magnitude and have values $\sqrt{3} \times$ the phase voltage. Applying these voltages to a delta-connected load as shown in Figure A.4 results in currents flowing from line to line through these loads. No neutral connection is involved. Looking at Junction 1 in Figure A.4 and applying Kirchhoff's current law, we see that $I_R + I_{BR} = I_{RY}$.

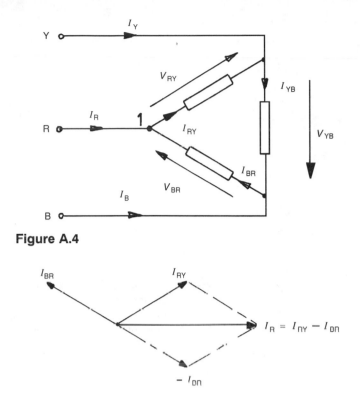

Figure A.4

Figure A.5

Transposing: $I_R = I_{RY} - I_{BR}$.

In Figure A.5, I_{RY} and I_{BR} have been chosen with equal magnitude and being in phase with their driving voltages. This means that the loads are resistive and the operating power factor is unity. Thus I_{RY} is parallel with V_{RY} in Figure A.3 and I_{BR} is parallel with V_{BR}. The operation $I_{RY} - I_{BR}$ involves reversing the phasor I_{BR} and the resulting addition gives I_R.

I_R by scaling is equal to $\sqrt{3}I_{RY}$ and equal to $\sqrt{3}I_{BR}$.

In a *balanced* delta-connected load $I_{LINE} = \sqrt{3}I_{PHASE}$.

In addition we see that the line current I_R is in the same direction; in phase with V_R in Figure A.1, i.e. the line current is in phase with the phase voltage of the supply generator. These results lead to the use of the formulae discussed in Chapter 1:

for either star or delta *balanced* loads, Power $= \sqrt{3}V_L I_L \cos\phi$
watts

However, if the loads are not balanced the above formula cannot be used. In Figure A.6 a load on the RY phase has been chosen which operates at unity power factor. I_{RY} is drawn parallel with V_{RY}. However, I_{BR} lags its voltage V_{BR} by an angle $\phi°$. It is flowing in an inductive load, and it does not have the same magnitude as I_{RY}.

$I_R = I_{RY} - I_{BR}$ as before.   (See Figure A.6)

We see that there is no obvious connection between the resulting magnitude of I_R and the two phase currents and the resulting phase angle θ is not equal to ϕ. With three different loads, performing the

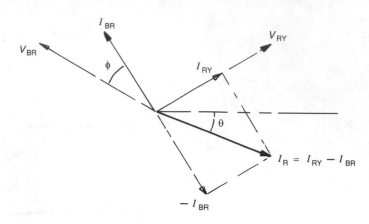

Figure A.6

phasor addition at each junction yields three different currents and angles. The balanced power formula cannot be used since there are three different values of I_L and three different angles.

Most delta-connected loads are in fact balanced, being typically three-phase motors or perhaps furnaces.

Answers

Chapter 1

Test your knowledge

1.1   4.01 A at an angle of $-78.35°$ from the red phase voltage.
1.2   (a) 32.8 A   (b) 18.94 A   (c) 869.4 A.
1.3   Current from 240 V source $= 47.86$ A.
      (a) min. P.D at point C $= 231.63$ V
      (b) Efficiency $= 0.96$ p.u.
1.4 $I_{AB} = 132.32$ A, $I_{BC} = 32.32$ A, $I_{CD} = 7.32$ A,
      $I_{ED} = 42.68$ A, $I_{AE} = 102.68$ A, $V_B = 242.06$ V,
      $V_C = 238.83$ V, $V_D = 238.46$ V, $V_E = 244.87$ V,
      Efficiency $= 0.97$ p.u.

Multiple choice questions

1 B;   2 C;   3 B;   4 A;   5 C, B;   6 C;
7 B,   8 B,   9 C,   10 C,   11 D;   12 C.

Chapter 2

Test your knowledge

2.1   (a) 0.015 m^2 (b) 22 turns
2.2   (a) $N_S = 6$ $I_S = 20$ A   (b) $V_S = 275$ V $I_P = 12.5$ A,
      (c) $N_P = 50$ turns, $V_P = 251$ V.
2.3   (a) $I_P = 0.625$ A $I_S = 12.5$ A,   (b) 0.96 Ω.
2.4   110 V section: 34.09 A, 440 V overhang 11.36 A.
2.5   (a) $0.0909 \times$ double-wound   (b) $0.943 \times$ double wound.
2.6   80 turns.
2.7   (a) 112.5 V   (b) 2.778 A   (c) 187.5 W.
2.8   (a) 1.27 A   (b) 0.273 A   (c) 65.52 W   (d) $I_{H+E} = 1.092$ A,
      $I_m = 5.08$ A.
2.9   (a) 11.92 A (b) (i) 13.45 A, (ii) $\cos 38.32 = 0.78$ (lagging).
2.10   Copper loss $= 1936.9$ W   Efficiency $= 0.915$ p.u.
2.11   (a) 27.18 A (b) efficiency $= 0.9825$ p.u. (Copper and core
      losses $= 96$ W).
2.12   (a) 8.02 Ω (b) 3.11 W   (c) 0.62 A.

Multiple choice questions

1 B;   2 C;   3 A;   4 D;   5 C;
6 A;   7 A, D, E;   8 B.

Chapter 3

Test your knowledge

3.1 (a) 180 V (b) 360 V, Power 14 400 W in both cases.
3.2 $\phi = 0.057$ Wb, $I_F = 0.715$ A.
3.3 Lap 504 V, 29.14 rev/s Wave 1008 V, 14.57 rev/s.
3.4 $E = 382.5$ V (a) 7.97 rev/s (b) $T = EI_A/2\pi n = 190.95$ Nm (c) 180.95 Nm.
3.5 (a) Total resistance = 4 Ω, added = 3.3 Ω (b) $I_A = 37.5$ A (c) $I_A = 150/3.3 = 45.5$ A.
3.6 (a) 19150 W (b) 0.837 p.u.
3.7 (a) 25832 W (b) 0.852.

Multiple choice questions

1 A; 2 C; 3 B; 4 C; 5 D; 6 B;
7 C; 8 C; 9 B; 10 A, D; 11 B.

Chapter 4

Test your knowledge

4.1 (a) 10 rev/s (b) 8 poles.
4.2 (1) 16.08 rev/s (2) 0.04 (4%), 2 Hz.
4.3 (a) 12.5 rev/s (b) 8 poles (four pole pairs).
4.4 Input to rotor = 9.5 kW, rotor loss = 380 W, torque = 90.72 Nm.
4.5 (a) $S = 0.2$ (b) $T = 80.7$ Nm (c) $T = 31$ Nm.
4.6 104 A, 24 Nm.
4.7 (a) 0.4 Ω (b) 64 Nm (c) 51.3 Nm (d) 50% (e) Stator input = 8766.7 W, Rotor output = 3783.4 W Efficiency = 43.2%.

Multiple choice questions

1 C; 2 C; 3 B; 4 B; 5 C; 6 B; 7 C; 8 B;
9 C; 10 C; 11 A; 12 B; 13 B, D; 14 B; 15 C.

Chapter 5

Test your knowledge

5.1 41.12 W.
5.2 (a) 5.5 A (b) No, cooling effect is not proportional to width
5.3 Fault current = 3873 A Total power $I^2R = 294.45$ kW; energy in 1/50 second = 6000 joules.
5.4 (a) 1.5 mm^2 (b) 2.5 mm^2 (c) 6 mm^2 (d) 1.5 mm^2 (Ambient temperature gives 14.08 A rating) (e) 2.5 mm^2.
5.5 (a) $3^2 \times 15 = 135$ W (b) $3 \times 20 = 60$ W.
5.6 $H = 500 \times 10^3$ A/m $B = 0.628$ webers per square metre (tesla) $\phi = 125.6\ \mu$Wb.

Multiple choice questions

1 C; 2 C; 3 B; 4 B; 5 A; 6 D;
7 B; 8 B; 9 C; 10 A; 11 C; 12 D.

Chapter 6

Test your knowledge

6.1   (a) 1765 lumens = 98 W (100 W practical)   (b) 140.4 cd
6.2   (a) 2948 cd   (b) 37 047 lumens.
6.3 $\dfrac{250}{500} \cos 63.4° + 2 = 2.224$ lux.
6.4   78.7°, 21 212 cd.
6.5   22 000 lumens = 8.6 tubes (9 practical) 3×3 arrangement.

Multiple choice questions

1 B;   2 B;   3 C;   4 C, E;   5 C;
6 B;   7 C;   8 A;   9 B;   10 A.

Chapter 7

Test your knowledge

7.1   (1) 54 A   (2) 1.67.
7.2   (a) 14 seconds (b) 1000 A (approximately due to scaling).

Multiple choice questions

1 C;   2 B;   3 A, C;
4 C, A;   5 C;   6 B;   7 B.

Chapter 8

Test your knowledge

8.1   (a) £44.5   (b) Average − 9.89p/kWh.
8.2   (1) £(9.5 + 100 + 412.5) = £522 Average 8.03p/kWh
     (2) £(13 + 82.5 + 105 + 240) = £440.5   Probably   worthwhile; costs of converting equipment?
8.3   MD = 228.3 kW, Average price 4.256p/kWh.
8.4   Original MD cost £14 285.7. New MD cost £11 111. Saving £3174.7. Capacitor charge £1607.6. Overall saving £1567.

Multiple choice questions

1 A, D;   2 C;   3 C;   4 B;   5 A;   6 A;
7 C;   8 C;   9 B, D;   10 C;   11 B.

Chapter 9

Multiple choice questions

1 B, G;   2 A (c), B (b), C (a);   3 C;
4 A, D;   5 D, F;   6 B.

Index